Leckie
the education publisher
for Scotland

Higher
APPLICATIONS
OF MATHS

Student Book

Craig Lowther, Bryn Jones

001/24112022

10 9 8 7 6 5 4 3 2

ISBN 9780008542290

Published by
Leckie
An imprint of HarperCollinsPublishers
Westerhill Road, Bishopbriggs, Glasgow, G64 2QT
T: 0844 576 8126 F: 0844 576 8131

leckiescotland@harpercollins.co.uk   www.leckiescotland.co.uk

HarperCollins Publishers
Macken House, 39/40 Mayor Street Upper, Dublin 1, D01 C9W8, Ireland

Publisher: Jenni Hall and Clare Souza
Author: Bryn Jones
Series Editor: Craig Lowther
Project manager: Peter Dennis

Special thanks to
Sarah Duxbury (cover design)
Andrew Duncan (proofreading)
Lynn McRobert (proofreading)
Ann Paganuzzi (illustrations)
SiliconChips (typesetting)

A CIP Catalogue record for this book is available from the British Library.

Acknowledgements
Whilst every effort has been made to trace the copyright holders, in cases where this has been unsuccessful, or if any have inadvertently been overlooked, the Publishers would gladly receive any information enabling them to rectify any error or omission at the first opportunity.

Printed in the UK.

CONTENTS

Introduction

About this book

This book provides a resource to practise and assess your understanding of the mathematics covered for the Higher Applications of Mathematics qualification. The content across the chapters builds to cover the skills specified in the Modelling, Statistics and Probability, Finance, and Planning and Decision Making areas of the course. Most of the chapters use the same features to help you progress. You will find a range of worked examples to show you how to tackle problems, and an extensive set of exercises to help you develop the whole range of operational and reasoning skills needed for your Higher assessment.

You should not necessarily work through the book from page 1 to the end. Your teacher will choose a range of topics throughout the school year and teach them in the order they think works best for you, as such you will use different parts of the book at different times of the year.

R Software

Throughout the textbook (and in the online resources) we have used the suite of R software tools and Microsoft Excel to store, manipulate and visualise data sets and calculations. By using R (or similar software) you will be able to analyse and interpret large data sets as part of the exercises in the book and for the project and final assessment. To aid you with your coding, this book presents code, diagrams and analysis as you will see them within RStudio.

Features

CHAPTER TITLE

The chapter title shows the operational skill covered in the chapter.

1 Modelling a Situation Mathematically

THIS CHAPTER WILL SHOW YOU HOW TO:

Each chapter opens with a list of topics covered in the chapter, and tells you what you should be able to do when you have worked your way through the chapter.

This chapter will show you how to:

- identify the appropriate variables in different contexts and solve Fermi problems
- recognise the shape and behaviour of linear, quadratic and exponential graphs and relationships

YOU SHOULD ALREADY KNOW:

After the list of topics covered in the chapter, there is a list of topics you should already know before you start the chapter. Some of these topics will have been covered before in Maths, and others will depend on having worked through different chapters in this book.

You should already know:

- how to find the volume of a sphere, cone, and prism
- how to find the surface area of a cylinder

EXAMPLE

Each new topic is demonstrated with at least one worked Example, which shows how to go about tackling the questions in the following Exercise. Each Example breaks the question and solution down into steps so you can see what calculations are involved, what kind of rearrangements are needed and how to work out the best way of answering the question.

Most Examples have comments, which help explain the processes. These comments are separated into three different types:

Select a strategy: what method are you adopting to tackle this problem?

Process data: the steps involved in working through the problem.

Interpret and communicate: the clear expression of the answer, including the correct units and decimal places.

Example 1.1

Estimate the number of times a typical person in Scotland will brush their teeth in their lifetime. State any assumptions you make.

You need to make two assumptions about the typical Scot here:

- their life expectancy
- how often they brush their teeth.

A reasonable estimate for the life expectancy of a Scot is 80 years. You can assume their teeth are brushed twice a day.

> Your dependent variable is how many times a typical person in Scotland brushes their teeth in their lifetime. Your assumptions about life expectancy and how often Scots brush their teeth daily are your independent variables.

Multiply the daily amount of brushing by 365 to find the yearly amount of brushing, then multiply by the life expectancy:

$2 \times 365 \times 80 = 58\,400$ brushes.

EXERCISE

The most important parts of the book are the Exercises. The questions in the Exercises are carefully graded in difficulty, so you should be developing your skills as you work through an Exercise. If you find the questions difficult, look back at the Example for ideas on what to do. Use Key questions, marked with a star, to assess how confident you feel about a topic.

Some exercises are designed to be completed on a computer and may require you to download files. All the files required for this textbook are provided online at

collins.co.uk/ScottishFreeResources/Maths

Exercise 1A

1 Estimate the number of Christmas crackers that are pulled in Scotland every December. State any assumptions you make.

2 Estimate the number of hours a typical person in Scotland spends eating in their lifetime. State any assumptions you make.

3 Amy drives a 5 minute commute to work. She decides to walk to work instead. Estimate the amount of money Amy would save on petrol over the course of one year, stating any assumptions you make.

KEY QUESTIONS

All exercises include Key questions, which are identified with a star ★. These questions are markers to help you assess your progress through the course.

DOWNLOAD

Many examples and exercise questions feature references to digital files. Where there is a file that can be downloaded, this is indicated by the following symbol: ⊕

ANSWERS

Answers to all Exercise questions are provided online at

collins.co.uk/ScottishFreeResources/Maths

PROJECT

Some chapters feature Project activities that are designed to help you think more deeply about the skills and techniques required for the Higher Applications project.

Think of a topic that interests you, and search online for a dataset about it. Make sure you choose a dataset which is a sample, not an entire population.

Answer the following questions:

- What is the dataset about?
- What do the column headings mean?

HINTS

Where appropriate, Hints are given in the text and in Examples, to help give extra support.

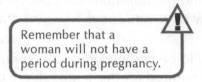

Remember that a woman will not have a period during pregnancy.

END-OF-CHAPTER SUMMARY

Each chapter closes with a summary of learning statements that show what you should be able to do when you complete the chapter. The summary identifies the Key questions for each learning statement. You can use the end-of-chapter summary and the Key questions to check you have a good understanding of the topics covered in the chapter.

- I can identify types of data: categorical, continuous numerical, discrete numerical. ★ Exercise 6A Q1
- I can explain the difference between a population and a sample. ★ Exercise 6A Q4
- I can identify biases introduced by non-random samples and find alternatives to avoid bias. ★ Exercise 6B Q2
- I can explain the influence of outliers on datasets. ★ Exercise 6C Q3, 5
- **I have created a research question for my project.**

1 Modelling a Situation Mathematically

This chapter will show you how to:

- identify the appropriate variables in different contexts and solve Fermi problems
- recognise the shape and behaviour of linear, quadratic and exponential graphs and relationships
- decide which formulae to use to model a given graph
- solve problems involving graphical models.

You should already know:

- how to find the volume of a sphere, cone, and prism
- how to find the surface area of a cylinder
- how the find the gradient of a straight line.

Mathematical models

A model is a simplified version of something that is more complicated. A map of Scotland is a model; it takes a complex thing like a country and simplifies it to some key information. A manager might rely on a mathematical model to understand the relationship between how much she spends on advertising and how many sales she makes.

A mathematical model uses mathematics to create an idealised, simplified version of a more complex situation.

Mathematical models can be used to make estimates about a context.

Fermi problems

Fermi problems are questions that can be difficult to answer precisely, but by making some sensible assumptions you can calculate a **reasonable estimate**. They call for simplified, 'back of the envelope' working out. The assumptions you make will either be common knowledge or something simple to look up.

Mathematical models have **independent** and **dependent** variables. Independent variables can be thought of as the **input** variables, which is the information you put into your model. The dependent variable is the **output** of your model. The output depends on the information you input.

Example 1.1

Estimate the number of times a typical person in Scotland will brush their teeth in their lifetime. State any assumptions you make.

You need to make two assumptions about the typical Scot here:

- their life expectancy
- how often they brush their teeth.

A reasonable estimate for the life expectancy of a Scot is 80 years. You can assume their teeth are brushed twice a day.

> Your dependent variable is how many times a typical person in Scotland brushes their teeth in their lifetime. Your assumptions about life expectancy and how often Scots brush their teeth daily are your independent variables.

Multiply the daily amount of brushing by 365 to find the yearly amount of brushing, then multiply by the life expectancy:

$2 \times 365 \times 80 = 58\,400$ brushes.

Example 1.2

Use the table here to estimate the number of bouncy balls that could fit into a hollow pyramid the size of the Great Pyramid of Giza. State any assumptions you make.

Pyramid	Height (metres)	Base length (metres)	Area of base (metres^2)
Khufu	146.60	230.45	53107
Khafra	143.87	215.16	46294
Giza	146.5	230	52900
Djoser	62	117.2	13736

Here are some assumptions:

- A typical bouncy ball has a radius of 2 cm.
- All the balls are identical.
- Spheres packed into pyramids will take up 70% of the volume.

You are not expected to know the optimal way to pack spheres into pyramids, so you can simply make sensible estimations. Spheres are not going to fit perfectly into a pyramid, and you have made an assumption that reflects this.

The volume of the pyramid can be calculated as:

$$V_P = \frac{1}{3}Ah = \frac{1}{3} \times (52900\,\text{m})^2 \times 146.5\,\text{m} \approx 2\,583\,283\,\text{m}^3$$

> A is the area of the base, h is the perpendicular height of the pyramid. You can find this information in the table.

The volume of one bouncy ball could be approximated as:

$$V_s = \frac{4}{3}\pi r^3 = \frac{4}{3} \times \pi \times 2^3 \approx 33.51\,\text{cm}^3$$

As you can assume that the balls can only fit into 70% of the volume of the pyramid, you can now calculate that:

$$70\% \times 2583283\,m^3 = 1808298.1\,m^3$$
$$\approx 1.8 \times 10^{12}\,cm^3$$

> You must convert metres cubed to centimetres cubed to have consistent units. Recall that 1 metre cubed is equal to 1 million centimetre cubed ($\times 100^3$).

Now you can complete the division:

Number of balls = volume available ÷ volume of one sphere
$$= 1.8 \times 10^{12}\,cm^3 \div 33.51\,cm^3 \approx 5.37 \times 10^{10}$$

So, approximately 50 billion bouncy balls.

Exercise 1A

1 Estimate the number of Christmas crackers that are pulled in Scotland every December. State any assumptions you make.

2 Estimate the number of hours a typical person in Scotland spends eating in their lifetime. State any assumptions you make.

3 Amy drives a 5 minute commute to work. She decides to walk to work instead. Estimate the amount of money Amy would save on petrol over the course of one year, stating any assumptions you make.

★ 4 On the radio programme *More or Less* it was estimated that the average number of periods a woman will have in her lifetime is around 350–450. Use all of the facts in the table to arrive at this estimate yourself.

> ⚠ Remember that a woman will not have a period during pregnancy.

Fact	Average
Average age to start having a period	12
Average age to stop having a period	51
Average number of periods a year	12
Average number of children a woman gives birth to	1.7

Are there any other facts that could improve your estimate?

5 A company wishes to put together a model to predict the number of sales their product makes, by considering the amount of money they spend on advertising. Describe what the independent and dependent variables would be in their model.

★ 6 A toy company manufactures steel pinballs for use in pinball machines. The average amount of pinballs they produce is 6000 per hour. The pinballs are spheres with a diameter of 2.5 cm. The company must buy steel in units of $10\,cm^3$. The steel units are processed and turned into balls using a grinder.

a Estimate the number of pinballs that the company produces per month. State any assumptions you make.

b Estimate the number of units of steel the company must buy per month, stating any assumptions you make.

7 A paint manufacturer produces 2000 litres of paint per hour. The paint is sold in steel cans with a cylindrical shape. The paint cans have a diameter of 10 cm and a height of 12 cm. The manufacturer must buy steel in sheets of 2 m by 2 m in order to produce the cans.

> ⚠️ Remember to consider wastage, not all of the steel squares will be able to be made into a cylinder.

a Estimate the amount of paint produced in a month. State any assumptions you make.

b Estimate the number of sheets, to the nearest 1000, that the paint manufacturer must buy per month. State any assumptions you make.

Recognising linear, quadratic and exponential graphs

You can use mathematical functions to create accurate models. In this course you will use **linear**, **quadratic** and **exponential** functions.

In a linear relationship, the dependent variable increases or decreases with the independent variable at a constant rate. If the dependent variable increases while the independent variable increases it is a positive linear relationship, if the dependent variable decreases while the independent variable increases it is a negative linear relationship. A linear graph is a straight-line graph.

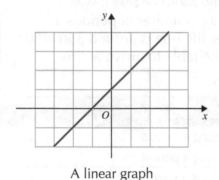

A linear graph

In a quadratic relationship dependent variables increase and then decrease, or decrease then increase. A quadratic graph looks like either a hill or a valley.

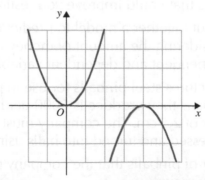

Quadratic relationships either decrease then increase (valley), or increase then decrease (hill)

An exponential relationship shows either growth that gets faster and faster, or decay that gets slower and slower.

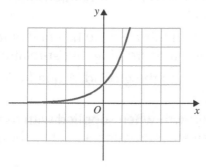

An exponential function showing ever increasing growth

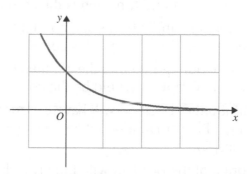

A function showing exponential decay

Example 1.3

Match the five graphs with the five contexts given in the table.

Context	Graph
The height of water in a bath as it is being filled.	
The height of a cylindrical candle after it is lit.	
The height of a ball as it thrown across a field.	
The number of bacteria in a Petri dish.	
The cost of a product that is being reduced by 10% each day.	

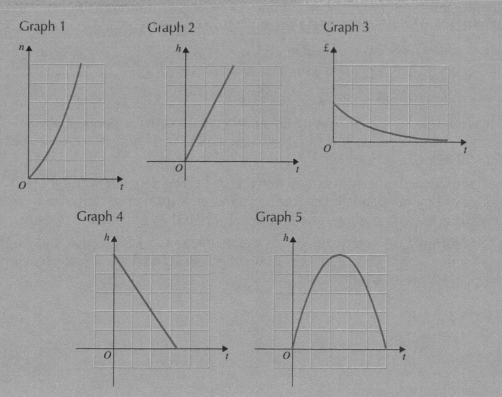

Graph 1

Graph 2

Graph 3

Graph 4

Graph 5

(continued)

Water enters a bathtub at a constant rate. As the amount of time that has passed increases, so does the height. Graph 2 shows a positive, linear relationship between height and time and is the correct match.

A cylindrical candle burns at a constant rate, but in this case the height would decrease over time. Graph 4 is correct as it shows a decreasing, linear relationship.

The ball would go up into the air, then back down to the ground. A quadratic graph, like Graph 5, would be the correct model.

The amount of bacteria will have exponential growth when supplied with food, starting with a small amount and then progressively increasing. Graph 1 shows this relationship.

The price of the product will initially reduce quickly, but as time passes the amount reduced will get progressively smaller. This is exponential decay. The price will never reach £0. Graph 3 shows this relationship.

Example 1.4

For each of the following formulae:

a Evaluate the formula at $x = 10$.

b Describe whether the dependent variable increases or decreases as the independent variable increases:

  i $y = -4x + 20$ ii $N = 2.3e^{2x}$ iii $P = 3x^2 - 4x + 2$

Evaluating the formula at a given value requires you to substitute the value into your formula:

  i $y = -4 \times 10 + 20$ $y = -20$

  ii $N = 2.3e^{2 \times 10} = 115\,879\,949$

  iii $P = 3 \times 10^2 - 4 \times 10 + 2$ $P = 262$

The first formula (i) is a decreasing linear relationship. The dependent variable (y) decreases as the independent variable (x) increases. You know it decreases because the coefficient of x is negative.

The second formula (ii) shows exponential growth. The dependent variable (N) increases as the independent variable (x) increases. Exponential formulae grow when the coefficient of x is positive, and they decay when the coefficient is negative.

The third formula (iii) is a quadratic. As the coefficient of x is positive, the graph will have a 'valley' shape: first P will decrease until it reaches a minimum point, and then it will increase as x increases.

One way to check your answers to these problems is to substitute lots of different values for x into the formulae using a spreadsheet, as you can see here:

C2	⌄	⋮	✕✓ _fx_	=2.3*EXP(2*A2)		

◢	A	B	C	D
1	x	y=–4x+20	N=2.3*e^(2x)	P=3x^2–4x+2
2	–5	40	0.00010442	97
3	–4	36	0.000771564	66
4	–3	32	0.00570113	41
5	–2	28	0.042125969	22
6	–1	24	0.311271151	9
7	0	20	2.3	2
8	1	16	16.99482903	1
9	2	12	125.5757451	6
10	3	8	927.886225	17
11	4	4	6856.20337	34
12	5	0	50660.87133	57
13	6	–4	374336.0203	86
14	7	–8	2765989.854	121
15	8	–12	20438054.2	162
16	9	–16	151017929	209

The spreadsheet shows the output of the three formulae at different values of x.

Exercise 1B

★ 1 For each of the following situations choose whether the best model would be linear, quadratic or exponential:

  a Predicting the number of ice-creams sold in an afternoon by considering the temperature.

  b How the value of new car that depreciates in value by 12% each year changes over time.

  c The population of people on Earth over time.

  d The height of an airplane during takeoff.

  e The height of a wooden rope bridge over a canyon.

2   The three graphs show how a population of rabbits varied over time.
State which graph could not model the rabbit's population.

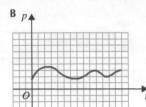

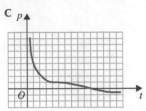

★ 3   Reindeer were transported to a small island with no native reindeer population.
The population of reindeer on the island was monitored over time. The graph
shows how the population, *p*, varies with time, years.

> This event happened on St. Matthew
> Island, when in 1944 the US coastguard
> introduced reindeer to the island.

a   Describe the relationship between population and
time between year 0 and year 22.

b   Why might the graph change after year 22?

c   In which years did the number of reindeer increase
the most?

0–11, 11–22 or 22–33

d   Estimate the reindeer population in year 11.

Island reindeer population

4   A model for population growth in bacteria uses the formula $P = 200e^{0.02t}$ where P
is the population and t is the time in minutes.

a   Use the model to estimate the population after 100 minutes.

b   Explain how the dependent variable changes as the independent variable
increases.

c   Explain why you should be cautious evaluating the formula at $t = 10\,080$.

★ 5   Using a spreadsheet, or otherwise, describe how the dependent variable varies as
the independent variable increases in the following formulae:

a $P = -x^2 + 40x - 2$

b $H = -3e^{0.01t} - 5$

c $S = 4 + 0.45W$

d $y = 2x^2 \times 2e^{-3x}$

Deciding which formulae to use to model a given graph and problem solving

Example 1.5

Consider this graph, which is missing a title and axis labels.

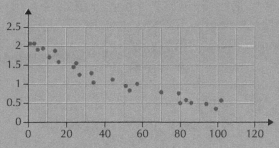

Which formula is the better fit for this model?

$P = 2e^{-0.02t}$ or $P = 1.9 - 0.02t$

One of those models is **exponential decay**, and the other is **a decreasing linear** relationship. Both formulae do a good job of modelling the graph, as you can see:

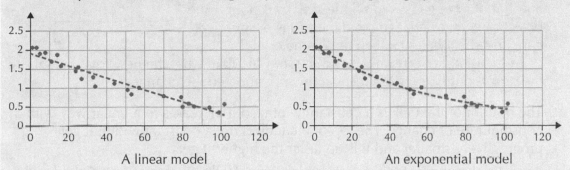

A linear model   An exponential model

If both models fit the data, how can you choose?

Understanding the situation that you are trying to model is important. Consider two different contexts that might produce the data you can see.

Scenario 1

In an experiment, patients were given an injection of a drug. Their blood was then tested to see what concentration of the drug remained after a period of time. The graph above shows the drug concentration levels on the vertical axis, and the amount of time that had passed on the horizontal axis.

The drug concentration decreases over time and eventually becomes zero. It is not possible to have a negative amount of a drug in your blood, and this is a strong hint that the exponential graph is more suitable.

Scenario 2

The lowest recorded temperature in a town is recorded each day. The graph above shows the temperature on the vertical axis, and time on the horizontal axis.

In this scenario it is possible for the downward trend to continue and become negative. In this case, a linear model is more appropriate because the exponential model will never give you a negative answer.

It is not enough to look at data, or graphs, by themselves to choose a model.

The model you fit must be appropriate to the context.

Example 1.6

A water tank is in the shape of a cylinder, with a spherical cap at the top. The cylinder has a height of 4 metres and a diameter of 5 metres. The total height of the tank is 5.5 metres, as shown in the diagram.

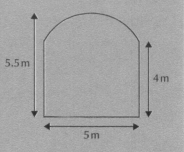

As the water tank fills, the height of the water level is recorded. The graph shows how the height of the water, h metres, varies with time, t hours.

a   Explain why the final part of the graph varies from the initial part.

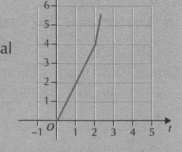

b h initially follows the model $h = mt + c$. Calculate the rate of change (by calculating the gradient) in the initial part of the graph, which is represented by m.

c   The designer of the water tank wants to increase the amount of time it takes to fill the tank by 3 hours. She does not want to change the dimensions of spherical cap, and she wishes to keep the diameter of the cylindrical part of the tank the same too. Calculate what the new total height of the tank would be if she successfully implemented the changes.

a   The bottom of the tank is a constant width, so it increases in a linear way with a constant rate of change. The top of the tank is not a constant width; it gets progressively smaller and fills up faster the higher it gets.

b   After 2 hours the water is 4 metres high. You also know that at time 0 the height is also zero:

Rate of change is $m = \dfrac{y_2 - y_1}{x_2 - x_1}$

$\dfrac{4\,m - 0\,m}{2\,h - 0\,h} = 2$ metres per hour.

> Remember to include units! As you are dividing metres by hours, the rate of change will be measured in *metres per hour*.

c   The tank takes 2 hours to fill up the cylindrical section, so the designer needs to increase this to 5 hours to meet her goal. You can substitute $t = 5$ into your equation:

$h = 2t + c$

c is zero as the y intercept is zero.

$h = 2 \times 5 = 10$

The new height of the cylindrical part of the tank must be 10 metres. You can see that the height of the spherical cap is 1.5 metres from the diagram, so the total height of the tank would be $10\,m + 1.5\,m = 11.5\,m$.

Exercise 1C

1   A clothing company measures the
    number of sales of swimming garments
    between September and April.

    Which formula would best model these
    results? Give a reason for your answer.

    A $S = -0.3d + 11\,000$

    B $S = (d - 106)^2 + 20$

    C $S = e^{-0.2x}$

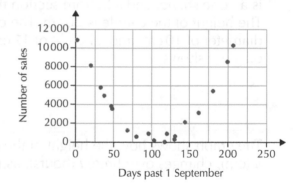

★ 2   A researcher produces a graph which
    shows the mass (tonnes) of vehicles plotted
    against the fuel consumption (miles per
    gallon) for that vehicle.

    a   Which formula would best model this
        graph? Give a reason for your answer.

        A $MPG = 37.285 - 5.344 \times mass$

        B $MPG = 800e^{-0.28 \times mass - 3}$

    b   Use your chosen model to predict the
        number of miles per gallon a vehicle
        with a mass of 4.5 tonnes could expect
        to have.

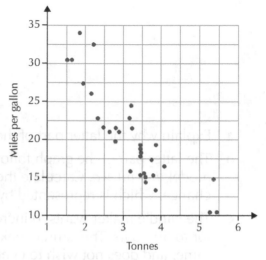

3   A bathtub is being filled up with water. The
    graph here shows the height of the water,
    h (cm), over time, t (minutes), when both
    taps are turned on.

    a   At what time does the bathtub
        overflow?

    b h initially follows the model $h = mt + c$.
        Calculate the rate of change in the
        initial part of the graph, which is
        represented by m.

    c   Sara fills the bathtub so that it is half
        full and then turns off the taps. She then
        turns on **one** of the taps. How long
        would it take before the bath reached
        its maximum height?

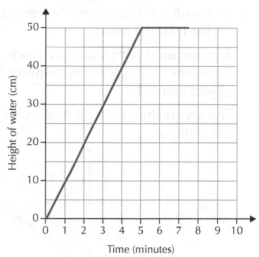

★ 4 A candle maker produces a candle with a top section that is a cone shape, and a bottom section that is a cylinder. The height of the candle is 20 cm. The cylinder has a diameter of 10 cm, and a height of 17 cm, as the diagram shows.

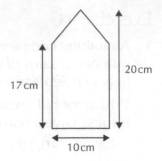

The graph shows how the height of the candle, h (cm), changes over time, t (hours), as it burns.

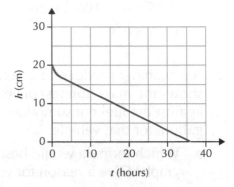

a Explain why the latter part of the graph is a straight line, rather than a curve.

b The latter part of the graph follows a linear model, $h = mt + c$. Calculate the rate of change, which is represented by m.

> Remember to include the units of m.

c The candle maker wants to increase the amount of the time the candle burns for to 50 hours. The candle maker wants the diameter of both parts to stay the same, and does not wish to change the shape of the cone at the top of the candle. What would the new, total height of the candle have to be so that it to burn for 50 hours?

5 A parachutist jumps out of a plane. In the initial jump phase, their velocity increases until it reaches 'terminal velocity', which is the maximum speed an object can fall.

The graph shows how the parachutist's velocity changes over time in the initial jump phase.

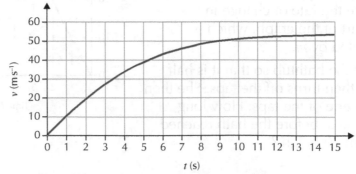

a What kind of mathematical model would be appropriate to model the initial jump phase? Give a reason for your answer.

b Which of the following would be a better model for the graph?

　i　$v = -55e^{-0.3t} + 52$

　ii　$v = -55e^{0.3t} + 52$

c Use your answer from part **b** to estimate the velocity of the parachutist at:

　i　3 seconds.

　ii　15 seconds.

　iii　300 seconds.

　Comment on your answers.

d The second stage of the jump takes place after the parachutist deploys their parachute. At this point, the parachutist falls at a constant rate. The height of the parachutist **during this second stage** can be modelled using $h = mt + c$, where h is the height in metres, m is the rate of change, and t is time in seconds.

　If the parachutist opens their parachute at 1500 m and then lands 3 minutes later, find the rate of change and the value of c.

- I can identify the appropriate variables in different contexts and solve Fermi problems. ★ Exercise 1A Q4, 6
- I can recognise the shape and behaviour of linear, quadratic and exponential graphs and relationships. ★ Exercise 1B Q1, 3, 5
- I can decide which formulae to use to model a given graph. ★ Exercise 1C Q2
- I can solve problems involving graphical models. ★ Exercise 1C Q4

2 Consistent Units of Measure

This chapter will show you how to:

- define appropriate units of measure for variables in a model
- check to see whether a model is consistent or inconsistent
- evaluate the output from mathematical models
- analyse mathematical models.

You should already know:

- how to substitute into formula, simplify an expression and rearrange an equation
- how to interpret a graph, such as a bar chart or a line graph
- how to find the area of simple polygons, such as trapezia and rectangles.

Consistent units of measure

Formulae need to have **consistent units** if they are to make sense and be useful as a model. Think about a simple example like speed:

$$\text{Speed} = \frac{\text{distance (miles)}}{\text{time (hours)}}$$

What unit would speed need to be measure in? In this example you are dividing miles by hours, so your unit for speed is $\frac{\text{miles}}{\text{hours}}$ (or simply miles per hour).

Example 2.1

A formula is given as:

$$T = \frac{2\pi r}{v}$$

where r is measured in metres, and v is measured in metres per second. Find the unit for T.

2 and π are constants so do not affect the units. So:

$$T = \text{metres} \div \frac{\text{metres}}{\text{seconds}} = \text{metres} \times \frac{\text{seconds}}{\text{metres}} = \text{seconds}$$

> $2\pi r$ is measured in metres, and it is being divided by v which is measured in metres per second. The metres cancel out and you are left with seconds.

Example 2.2

Here is a graph showing the force being placed on an object over time:

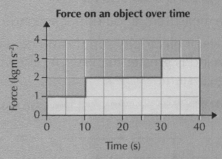

Force on an object over time

Find the area of the shaded region underneath the graph. Give a suitable unit for your answer.

You can think of the area underneath this graph as three rectangles, labelled A, B and C.

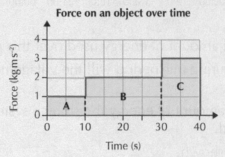

Force on an object over time

$J = A + B + C = 1\,kg \times m \times s^{-2} \times 10\,s + 2\,kg \times m \times s^{-2} \times 20\,s + 3\,kg \times m \times s^{-2} \times 10\,s$

$= 10\,kg \times m \times s^{-1} + 40\,s + m \times s^{-1} + 30\,kg \times m \times s^{-1}$

$= 80\,kg\,m\,s^{-1}$

$kg \times m \times s^{-2} \times s$ is equal to $kg \times m \times s^{-1}$.

Exercise 2A

1 A formula is given as: $\rho = \dfrac{m}{v}$ where m is measured in kilograms and v is measured in cubic metres. Find the unit for ρ.

2 Pressure (p) can be found using the formula $p = \dfrac{F}{A}$ where F is measured in Newtons (N) and A is an area measured in square metres. In which unit can you measure pressure?

★ 3 A model is given as: $I_P = \dfrac{mgrt^2}{4\pi^2}$ where m is measured in kilograms, g is measured is metres per second squared ($\frac{m}{s^2}$), r in metres and t in seconds. Show that the unit for I_P is $kg \times m^2$.

4   If $J = N \times m$ (where J is a joule, N is a Newton and m is a metre), and $N = kg \times \dfrac{m}{s^2}$ (where kg is a kilogram and s is a second), and $W = \dfrac{kg \times m^2}{s^3}$ (where W is a watt), show that a watt can be measured in joules per second.

★ 5   The graph shows the rates of electricity supply and demand over a 24-hour period for a village. When demand is greater than supply, the village needs to retrieve electricity from a storage system.

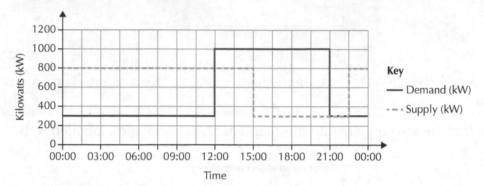

a   Determine the total amount of energy used over this period.

b   How many hours during this period will the village need to make use of their storage system?

c   Find the minimum amount of energy the storage system must be able to store to meet the demand.

6   A speed–time graph is shown.

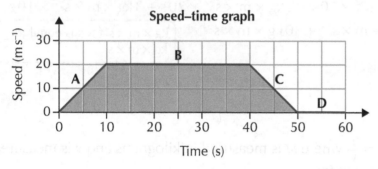

a   During which section of the graph was the object's speed decreasing: A, B, C or D?

b   During which section was the object stationary?

c   Find the area of the shaded region, underneath the graph, giving appropriate units. What does this tell you about the object?

Evaluating the output of mathematical models

When presented with the output of a mathematical model, you can consider how accurate or plausible it is. You should consider whether:

- the model uses consistent units – a formula that has *inconsistent* units cannot be a valid model
- the statements being made are **accurate** to the level of **precision** given
- the data are being **extrapolated** beyond what would be valid for the model
- the claims being made are plausible or reasonable

> A statement is **accurate** if it is factually true. Precision is how much certainty you have about a figure. For example, "My height is between 1 metre and 5 metres" is an accurate statement, but not a precise one. "My height is 3.235 42 metres" is a very precise statement, but not at all accurate.

Example 2.3

Andrew says: "I weigh 70 kg, and my height is 170 cm. That means my height is greater than my weight."

Is Andrew correct?

Andrew's comparison is not valid, because he is using inconsistent units. It is not meaningful to compare a kilogram to a centimetre. If Andrew had instead said "I weigh 70 000 g, and my height is 1.7 m. So my weight is greater than my height." then this would be the opposite of his first statement with the same height and weight! It's important for units to match when you make comparisons.

Example 2.4

Here is snippet from a news article:

> In 1950 there were 35 488 231 hedgehogs in the UK. Today, the hedgehog population is less than 1 million. At the current rate of decline the species will be extinct in the UK within the next 20 years.

Evaluate the news snippet.

The figure for the hedgehog population in 1950 is much too precise. It is very unlikely that an exact value for hedgehogs can be given.

Data are being extrapolated beyond what is reasonable; it is unlikely that the decline will continue at "the current rate". This newspaper is assuming a linear model, but it's plausible that the population will decline, for example, as an exponential decay.

Exercise 2B

1 Ella draws a square with side lengths of 4 cm. She says: "The perimeter and the area of my square are equal." Comment on what Ella says.

★ 2 A news reporter says: "Scotland's gross domestic product (GDP) is around £150 billion a year, and our debt is around £220 billion. An independent Scotland's debt would be worth more than its economy." Is this a valid statement?

3 A student is observing how octopuses can fit through small, circular holes. She produces the following model:

maximum hole radius (cm) = octopus length (cm) × mass (kg) × 2.349

Is this a valid model? Explain your reasoning.

★ 4 Below is a press release for an e-cigarette company:

"343 602 Scots reported regularly vaping in 2022, up from 130 240 in 2018. This suggests that by 2035, the number of vaping adults in Scotland will exceed 10 million." Evaluate this statement.

★ 5 a During the COVID-19 pandemic, countries tried to keep records of the number of deaths related to the illness. One model used "excess deaths", in which the number of people who died in a country was compared with the number of deaths normally recorded in a year. Below is a table printed in a newspaper showing the excess deaths for some countries.

Country	Excess deaths (per 100 000 people)
China	Between 100.24 and 150
UK	Between 200 and 220
New Zealand	Between −50 and −60
India	Between 60 and 650

Which country's data are the most precise?

b Emilie says: "The newspaper has made a mistake with New Zealand's figures. It suggests that the pandemic has brought people back to life!" Comment on what Emilie says.

★ 6 Some conservationists propose reintroducing lynx into the Highlands. Lynx prey on deer, and deer currently have no natural predators. A scientist is considering how the population of lynx might change if it was re-introduced. His model looks like this:

$$\text{Population} = 10e^{0.2t}$$

a Explain how the population will change over time (t).

b Evaluate this model, suggesting a variable that could be added to make the model more accurate.

7   One way to measure fuel efficiency in cars is by using the unit "litres per kilometres". Here is a table of information about the journeys of different cars.

Car	Distanced travelled (km)	Amount of petrol used (litres)
A	400	20
B	600	35
C	125	10
D	80	15

a   For each vehicle calculate the fuel efficiency in litres per kilometres.

b   Show that the unit "litres per kilometres" is equivalent to "mm^2".

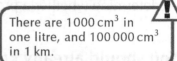

There are $1000\,cm^3$ in one litre, and $100\,000\,cm^3$ in 1 km.

c   Here is a table showing some information about four cylinders.

Cylinder	Total length (km)	Volume (*l*)
A	400	20
B	600	35
C	125	10
D	80	15

Find the area of the circular cross-section of the cylinder in each case.

d   Gareth says: "If you were to take all the petrol that you burn on a car journey (in litres) and put it into a long tube that was the exact length of the distance you drove (in km), then the area of the circular cross section of that tube would be equal to the fuel efficiency of the car (in litres per kilometres)." Is this a valid model?

• I can identify appropriate units of measure for variables in a model. ★ Exercise 2A Q3, 5
• I can identify where a model is consistent or inconsistent. ★ Exercise 2B Q2
• I can evaluate the output from mathematical models. ★ Exercise 2B Q5
• I can analyse mathematical models. ★ Exercise 2B Q6

3 Errors and Tolerance

This chapter will show you how to:

- use and understand tolerance notation
- convert between absolute and relative errors
- calculate or estimate the absolute and relative errors in a model
- determine whether a process is within tolerance.

You should already know:

- how to interpret tolerance notation of the form 50 ± 0.05
- how to re-arrange linear equations and formulae, for example making a the subject of the formula $v = u + at$
- the definition of independent and dependent variables
- how to find the average (mean) of a dataset
- how to convert between percentages and decimals.

Types of error

When working with mathematical models, you often estimate values or work to only a certain level of **precision**. A confectioner may intend to produce packets of sweets containing 20 sweets each, but in reality some packets may contain more or fewer than that. You call this uncertainty the **error**.

The **absolute error** is the actual value of the error (including the units). The absolute error is always positive. For example, 20 ± 3 sweets tells you the number of sweets could be as low as 17, or as high as 23. The absolute error is 3. You denote the absolute error as Δx.

> Δx is pronounced "delta ex"; the small triangle is the capital Greek letter delta.

The **relative error** is the absolute error divided by the intended measurement and is often expressed as a percentage. Denote your intended measurements as x_0, which refers to what your measurement would be if there was no error at all. The relative error is denoted as $\frac{\Delta x}{x_0}$. In the sweets example, the relative error can be found using $\frac{3}{20} = 15\%$, so the number of sweets per packet is $20 \pm 15\%$.

Example 3.1

A circle is drawn with a circumference of 50 ± 2 cm. Calculate the relative error.

$$\frac{\Delta x}{x_0} = \frac{2\,\text{cm}}{50\,\text{cm}} = 4\%$$

> The absolute error, Δx, is 2 cm. The intended measurement, x_0, is 50 cm.

Example 3.2

A weighing scale measures a person's mass in kilograms with an error of 5%. Callan weighs himself as 50 kg. What is the maximum mass and minimum mass of Callan?

$50 \text{ kg} \times \pm 0.05 = \pm 2.5 \text{ kg}$

So Callan is at most $50 \text{ kg} + 2.5 \text{ kg} = 52.5 \text{ kg}$ or at least $50 \text{ kg} - 2.5 \text{ kg} = 47.5 \text{ kg}$.

> You multiply by the relative error to find the absolute error.

Exercise 3A

1 A baker produces cakes advertised as weighing 400 g. To comply with the law, the cakes sold must be within 12 g of the advertised mass.

   a What is the value of Δx?

   b What is the value of x_0?

   c Calculate the relative error.

2 Cakes sold as weighing 1000 g can have a relative error of 1.5%. Express this as an absolute error and state the maximum and minimum mass the cakes could be sold for.

★ 3 The table sets out the tolerable errors that bakers should adhere to when baking bread.

Mass (g)	Tolerable error
200–300	9 g
301–500	3% of the mass
501–1000	15 g
1001–10 000	1.5% of the mass

Answer the following:

   a Calculate the relative error for a loaf weighing 250 g.

   b Calculate the absolute error for a loaf weighing 1250 g.

   c A baker makes a batch of 400 g loaves of bread. She takes a sample of 5 loaves and weighs them. Their masses are: 403 g, 410 g, 389 g, 388 g, 411 g. For the batch to meet the required standard, all the samples must be within the tolerable error, and the mean of the samples must be within 1% of the intended mass. Does the batch meet the standard?

4 Crab fishing in Scotland is regulated to avoid juvenile crabs being harvested. Fishing juvenile crabs has a negative impact on the sustainability of crab fishing. Fisheries have a minimum landing size for crabs, which is the smallest a crab can be before it can be harvested. Some information is given in the box:

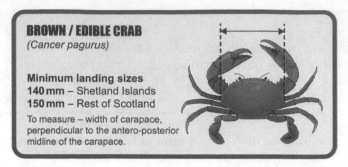

BROWN / EDIBLE CRAB
(Cancer pagurus)

Minimum landing sizes
140 mm – Shetland Islands
150 mm – Rest of Scotland
To measure – width of carapace, perpendicular to the antero-posterior midline of the carapace.

Richard is working in the Shetland Islands. Richard measures the brown crabs he catches using an absolute error of 2 mm.

a Calculate the relative error of a brown crab that Richard measures as 179 mm.

b What is the smallest measurement Richard could take for the landing size of a crab and be confident that the legal minimum landing size had been met?

Ross is working in the Firth of Forth. Ross measures brown crabs with a relative error of 5%. Ross measures a brown crab as 155 mm.

c Calculate the absolute error of this measurement.

d Can Ross be confident that it meets the minimum landing size?

Errors in a model

In a model, if there are errors in the independent (input) variables there is likely to be an error in the dependent (output) variable. For example, if you measure the radius of a circle and make an error in that measurement, when you then calculate the area of that circle you will have an error in your area.

Suppose you measure the radius of a circle as 10 cm, but you have an error of ±20%. The true radius could be as large as 10 cm × 1.2 = 12 cm. When calculating the area, a circle with a radius of 10 cm has an area of $100\pi\,cm^2$. A circle of radius 12 cm has an area of $144\pi\,cm^2$. The absolute error in the area is $44\pi\,cm^2$, a 44% relative error.

Power Law

When working out errors, there is a useful rule called the Power Law. For a model $z = ax^n$, the relative error of z is approximately equal to the relative error of x multiplied by its power. For example, in the model $z = 3x^4$, if x is correct to a relative error of 5%, then the relative error of z is approximately $4 \times 5\% = 20\%$.

Using the example of the circle, $A = \pi r^2$, the relative error of your radius calculation is 20%, you can estimate that the relative error of the area as $2 \times 20\% = 40\%$.

In models of the form $z = ax$ (linear) and $z = \frac{a}{x}(x^{-1})$, the relative error of x is equal to the relative error of z. When your model is of these forms, the relative errors in the independent and dependent variables are the same.

Estimating errors

When independent variables are being multiplied or divided, you can **estimate** the relative error of the dependent variable by **adding** the relative errors of each independent variable.

Example 3.3

Matthew measured the length, height and width of a cuboid. His relative errors in measuring are shown in the table.

Measurement	Relative error (%)
Length	5
Height	4
Width	7

Matthew is finding the volume of the cuboid. Estimate the combined relative error of Matthew's volume calculation.

Volume is found by multiplying variables: $V = l \times w \times h$.

You can estimate the combined relative error as $\frac{\Delta V}{V_0} \approx 5\% + 4\% + 7\% = 16\%$.

Example 3.4

A sprinter runs a distance while being timed with a stopwatch. The distance of the track has been measured with a relative error of 0.5%. The stopwatch measures the sprinter with a relative error of 3%. Estimate the relative error of the sprinter's speed.

Speed is found by dividing distance by time. $S = \frac{D}{T}$.

You can estimate the relative error as $\frac{\Delta S}{S_0} \approx 0.5\% + 3\% = 3.5\%$.

Discrete objects

Discrete objects are things that can only be counted as integers. For example, the number of books in a library, or the number of people in a room. When measuring the number of discrete objects, you will always find that the answer is a whole number amount.

Suppose the average number of students in a high school classroom was 30.2 students. If you chose to enter a classroom at random, and count the number of students, you would never find a classroom with exactly 30.2 students. Students are discrete objects, so whenever they are counted they will always be measured as an integer. This means that when counting discrete objects to estimate an average, an error can be introduced.

The error when counting discrete objects is ±0.5. If you entered a classroom at random and counted 30 pupils, you may decide the estimate of the average number of students is between 29.5 and 30.5.

Example 3.5

A biologist is estimating the number of deer in a woodland. He surveys a kilometre square of forest and counts 17 deer. The area of woodland being surveyed is a rectangle that is 7 km long, and 5 km wide. The **width** of the woodland has been estimated within an error of ± 0.1 km.

Estimate the number of deer in the woodland area, giving an estimate of the relative error of your calculation.

You can estimate the number of deer using:

$P = 17$ deer per $km^2 \times (7 \, km \times 5 \, km) = 595$ deer.

You can work out the relative error for the width calculation:

$$\frac{\Delta W}{W_0} = \frac{0.1 \, km}{5 \, km} = 2\%.$$

> The average number of deer per square kilometre might not be an integer. When the biologist counts how many deer he sees in his surveyed area, he will only count a whole number of deer and this might introduce an error.

What is the relative error of the number of deer counted? Deer are discrete objects so you can estimate the error in their density as ±0.5.

The relative error of deer density then is $\frac{\Delta D}{D_0} = \frac{0.5}{17} = 2.9\%$.

As the model is a product ($P = D \times L \times W$) you can sum the relative errors to estimate the relative error of P. $\frac{\Delta P}{P_0} \approx 2\% + 2.9\% = 4.9\%$.

So the estimated number of deer in the woodland is 595 deer ± 4.9%.

Exercise 3B

1   Scott is measuring a triangle to calculate its area. He measures the base as being 4 cm (with an error of 6%) and the perpendicular height as 10 cm (with an error of 5%). What is the relative error of his area calculation?

2   Density is calculated using the formula $\rho = \frac{m}{V}$ where m is measured in kilograms, and V in cm^3. The table shows the measurements of different liquids and the relative errors in measuring them. For each liquid calculate the density, giving an estimate of the relative error of your calculation.

Liquid	Mass (kg)	Mass error (%)	Volume (cm³)	Volume error (%)
Camelina oil	200	5	0.22	2
Petrol	500	2	0.68	6
Water	250	0.45	0.25	0.55
Methane	400	11	0.86	2

★ 3 The health of a coral reef can be measured by the density of sharks that live within the environment, measured in sharks per km^2. A team of marine biologists are conducting a biological survey in a coral reef. They are investigating life in an area of the sea that is 30 km long by 40 km wide. The 40 km measurement has been estimated within an error of ±0.2 km. The team survey a 1 km^2 area of water and count 24 sharks.

   a Estimate the number of sharks in the total area of sea, including an estimate of the relative error of your calculation.

   b What is the minimum number of sharks you would expect to find in the whole section of sea?

4 Oliver wants to know whether there is any interest in starting a rugby club at his school. He decides to survey a random tutor group to estimate the number of people who would be interested in joining. In the tutor group 2 people say they are interested. Oliver estimates that there are 40 tutor groups in his school (within an error of ±4 groups). Estimate the number of students who may be interested in joining the club, including an estimate of the **absolute error** of your calculation. Comment on your result.

Calculating errors

If you have a model where variables are being added or subtracted, you can work out the **absolute error** of the **dependent variable** by summing the **absolute errors** of the **independent variables**, taking into account any coefficients.

Example 3.6

The perimeter of a rectangle is given by $P = 2l + 2w$. If there is an **absolute error** of 3 cm in the measurement of the length l and an **absolute error** of 4 cm in width w, find the absolute error in the perimeter P.

The absolute error in P is found as:

$\Delta P = 2 \times \Delta l + 2 \times \Delta w = 6\,cm + 8\,cm = 14\,cm.$

You can check this by considering a specific example.

> Remember that delta l and delta w are the absolute errors of l and w. You multiply by 2 because these variables are being multiplied by 2 in the model.

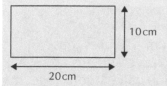

10 cm

20 cm

(continued)

In this rectangle the maximum error you could make in finding the perimeter of the rectangle would be if you measured the 10 cm length as 13 cm, and the 20 cm length as 24 cm. That would give a perimeter of the rectangle as:

$P = 2 \times 13\,\text{cm} + 2 \times 24\,\text{cm} = 74\,\text{cm}$.

The actual perimeter of the rectangle is 60 cm. Your calculation showing the difference was 14 cm is the same as the absolute error P you calculated earlier. This error will be the same no matter what dimensions of rectangle you chose.

Example 3.7

A model is given as $z = 3a - 5b$. The **absolute error** in a is 4 kg, the **absolute error** in b is 6 kg. Find the absolute error in z.

The absolute error in z is found as:

$\Delta z = 3 \times \Delta a + 5 \times \Delta b = 3 \times 4\,\text{kg} + 5 \times 6\,\text{kg} = 12\,\text{kg} + 30\,\text{kg} = 42\,\text{kg}$.

The important thing to remember is that you are **adding** the absolute errors, taking into account the coefficients, even though the model uses subtraction.

Example 3.8

Enoch measures the radius of a circle with a relative error of 5% and then calculates the circumference of the circle. What is the relative error of his circumference calculation?

The circumference of a circle is found using $C = 2\pi \times r$, where r is the radius of the circle. This is a linear model, so the relative error of the circumference will also be 5%.

You can check this by picking a random number for Enoch's measurement.

If $r_0 = 5\,\text{cm}$ then $\Delta r = 5\% \times 5\,\text{cm} = 0.25\,\text{cm}$.

$C_0 = 2\pi \times 5\,\text{cm} = 10\pi\,\text{cm}$

$C = 2\pi \times 5.25\,\text{cm} = 10.5\pi\,\text{cm}$

In this case, the absolute error (ΔC) would be:

$10.5\pi\,\text{cm} - 10\pi\,\text{cm} = 0.5\pi\,\text{cm}$.

So the relative error would be:

$\dfrac{\Delta C}{C_0} = \dfrac{0.5\pi\,\text{cm}}{10\pi\,\text{cm}} = 0.05 = 5\%$.

Both the π and the cm cancel in the fraction.

Example 3.9

Oliver measures the lengths of a cuboid as 3 cm, 4 cm and 5 cm. He uses these measurements to find the volume of the cuboid. In fact, the correct lengths are 3.2 cm, 3.9 cm and 5.3 cm. Find the relative error of his volume calculation.

$V = l \times w \times h$

$V_0 = 3.2\,\text{cm} \times 3.9\,\text{cm} \times 5.3\,\text{cm} = 66.144\,\text{cm}^3$

$V = 3\,\text{cm} \times 3\,\text{cm} \times 5\,\text{cm} = 60\,\text{cm}^3$

$\Delta V = 66.144\,\text{cm}^3 - 60\,\text{cm}^3 = 6.144\,\text{cm}^3$

$\dfrac{\Delta V}{V_0} = \dfrac{6.144\,\text{cm}^3}{66.144\,\text{cm}^3} = 9.3\%$

> You use the subscript zero to indicate the result when there is no error.

> The absolute error is the difference between the correct measurement and the measurement as measured with the error. It is always positive.

Exercise 3C

1 A model is given as $z = 5a - 2b$. The absolute error in a is 3 cm, the absolute error in b is 2 cm. Find the absolute error in z.

2 A model for the perimeter of a rectangle is given by:
   $P = 2l + 2w$. Suppose you know the length (l) is 5 cm ± 1% and the width (w) is 10 cm ± 1.5%. Calculate the relative error of P.

> Find the absolute errors in l and w.

> A common error is to think the answer should be 1.5% + 1% = 2.5%. (Or 5% if you doubled the relative errors). The mistake is confusing absolute errors, which can be summed in models with addition, with relative errors that cannot be summed here.

3 Mirren measures the length of a square as 5 cm, **to the nearest cm**.

   a What is the relative error of her length measurement?

   b Mirren uses her measurement to find the area of the square. What is the relative error of her area calculation?

★ 4 Bethany measures the lengths of a cuboid as 5 cm, 7 cm and 10 cm. She uses these measurements to find the volume of the cuboid. In fact, the correct lengths are 5.4 cm, 7.1 cm and 9.8 cm. Find the relative error of her volume calculation.

5 A printing press can print 30 000 newspapers per hour, with a relative error of 11%. The printer runs for 210 minutes overnight.

   a Write down a formula stating the relationship between time T (expressed in hours), the printing rate R (expressed in papers per hour), and the number of newspapers printed overnight, N. Make N the subject of your formula.

   b Calculate N_0.

   c Calculate the absolute error of N.

   d Write down the relative error of N.

   e What is the minimum number of papers printed overnight?

6 A paperboy delivers 4 newspapers a minute, with a relative error of 20%. He has 420 papers to deliver.

   a Write down a formula stating the relationship between the number of papers N, the delivery rate R (expressed in papers per hour), and the time to deliver the papers T (expressed in hours). Make T the subject of your formula.

   b Write down the relative error of T.

c   Calculate T_0.

d   Calculate the absolute error of T.

e   The paperboy starts his round at 6 am. He must finish his round before 8 am to be on time for a dentist appointment. Is it possible he will miss his appointment?

★ 7   A solar panel generates 400 watts of electricity per hour when exposed to direct sunlight. The relative error in this rate is 15%. Today the solar panel was exposed to direct sunlight for 10 hours. Calculate the total amount of electricity generated today, and state the absolute error of your result.

8   A quadratic model is given as $y = x^2 - 11x + 24$. x is measured as 9 with a relative error of 2%. Calculate the relative error of y.

Tolerance

So far, you have been calculating and estimating the errors in the dependent variable. But often you will want to be able to work backwards. A **tolerance** is the maximum error you are prepared to accept in the dependent variable.

When you measure things, you often do so to use your measurements in a calculation. For example, a student committee organising a school prom may want to know how much money, approximately, they will have to hire a venue. The committee will want to have an approximation, but also know that the approximation has a reasonable tolerance. One way to estimate how much money will be available is to multiply the estimated number of people attending, with the estimated price of a ticket.

It would be natural for the committee to ask: "How good do your estimates for the ticket price and the number of attendees have to be so that your venue budget calculation is correct within a tolerance of 10%?"

Example 3.10

Heather is drawing a circle and wants the area to be $20\,cm^2$. The **tolerance** for the area is 2%. Calculate the maximum absolute and relative error of Heather's radius measurement.

There are two methods you can use to answer this question:

Method 1

The model is $A = \pi r^2$. You can rearrange to make r the subject: $r = \sqrt{\dfrac{A}{\pi}}$.

The intended measurement would be:

$r_0 = \sqrt{\dfrac{20\,cm^2}{\pi}} \approx 2.5231\,cm.$

The maximum area would be $20\,cm^2 \times 102\% = 20.4\,cm^2$.

The minimum area size would be $20\,cm^2 \times 98\% = 19.6\,cm^2$.

This means the maximum radius size is

$$r = \sqrt{\frac{20.4\,\text{cm}^2}{\pi}} \approx 2.5482\,\text{cm}.$$

The minimum radius size is

$$r = \sqrt{\frac{19.6\,\text{cm}^2}{\pi}} \approx 2.4978\,\text{cm}.$$

The **absolute error** in r is:

either: $\Delta r = r - r_0 = 2.5482\,\text{cm} - 2.5231\,\text{cm} = 0.0251\,\text{cm}$

or: $\Delta r = r - r_0 = 2.5231\,\text{cm} - 2.4978\,\text{cm} = 0.0253\,\text{cm}.$

> The absolute error is always positive.

You are considering the worst-case scenario, so you can choose the smallest absolute error for r. This is how close Heather must draw the radius to be confident that the area is within tolerance.

The **relative error** is:

$$\frac{\Delta r}{r_0} = \frac{0.0251\,\text{cm}}{2.5231\,\text{cm}} \approx 1\%.$$

Method 2

Rearrange $A = \pi r^2$ to $r = \sqrt{\dfrac{A}{\pi}} = \left(\dfrac{A}{\pi}\right)^{\frac{1}{2}}$. As the index of A is one half, the relative error of r is:

> You are multiplying by $\frac{1}{2}$ because the index of A is one-half.

$$\frac{\Delta r}{r_0} = 2\% \times \frac{1}{2} = 1\%$$

As with method 1, you know that $r_0 = \sqrt{\dfrac{20\,\text{cm}^2}{\pi}} \approx 2.5231\,\text{cm}.$

So the absolute error of r is $2.5231\,\text{cm} \times 1\% = 0.0253\,\text{cm}$

Remembering the Power law will save you time.

Exercise 3D

1 In a woodwork lesson, Cameron is asked to cut a circular piece of wood that has a **circumference** of 15 cm. The **tolerance** for the circumference is 10%. Calculate the maximum absolute error of Cameron's radius measurement.

★ 2 A machine produces bouncy balls in the shape of a sphere. The balls must have a volume of $35\,\text{cm}^3 \pm 9\%$. Calculate the absolute error of the radius measurement.

3 A model is given as $A = 3b^4$. If you want A to equal $100\,\text{m}^4$ within a tolerance of 24%, calculate the relative error of the measurement of b.

★ 4 Compound interest is calculated using the model $A = P(1 + i)^n$. P is the amount invested, i is the annual interest rate, and n is the number of years. Dylan knows that he wants to invest his money in a bank account that pays 4% interest per year, and that he will leave the money in the account for 5 years. After 5 years Dylan will need £7000 to afford to go on his dream holiday. There is a tolerance of 10% on this £7000 target. Calculate the relative error of P, the amount to be invested. State the minimum amount Dylan needs to invest.

- I can interpret tolerance notation, and convert between absolute errors and relative errors. ★ Exercise 3A Q3
- I can calculate the absolute and relative errors in a model. ★ Exercise 3C Q4, 7
- I can estimate the absolute and relative errors in a model. ★ Exercise 3B Q3
- I can calculate the tolerance of a model and determine whether a process is within tolerance. ★ Exercise 3D Q2, 4

4 Spreadsheet Skills

This chapter will show you how to:

- sort and organise data in a spreadsheet
- extract basic statistical information and produce charts from data
- use the following spreadsheet functions: **MIN, MAX, AVERAGE, MEDIAN, STDEV, PEARSON, SUM, PRODUCT, IF, AND, OR, ROUND, INT, COUNTIF, GOAL SEEK**
- implement recurrence relations in a spreadsheet.

You should already know:

- how to calculate the median, mean, standard deviation and semi-interquartile range of a given dataset
- how to interpret scatter plots
- how to calculate compound interest.

Sorting and organising data

Spreadsheets can contain large amounts of data. Before you analyse data, you may have to sort and organise them to make things easier.

Statistical functions

MIN() – returns the minimum number

MAX() – returns the maximum number

AVERAGE() – returns the mean

MEDIAN() – returns the median

STDEV() OR **STDEV.S()** – returns the standard deviation given a sample

PEARSON() – returns the correlation coefficient

Example 4.1

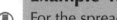

For the spreadsheet named "4.1 Student Data":

a   Complete the "Pupil Number" column.

b   Sort the class in order of youngest pupil to oldest pupil.

c   Calculate the **correlation coefficient** between age and test score, and create a suitable plot.

d   Find the mean, median, and standard deviation for class A and class B. Make two valid comparisons between the two classes.

a You do not have to manually type the numbers for the students. The spreadsheet software will spot a pattern and continue it. Select the 1 and 2, click the black square in the corner of the cell, and drag down to fill in the cells.

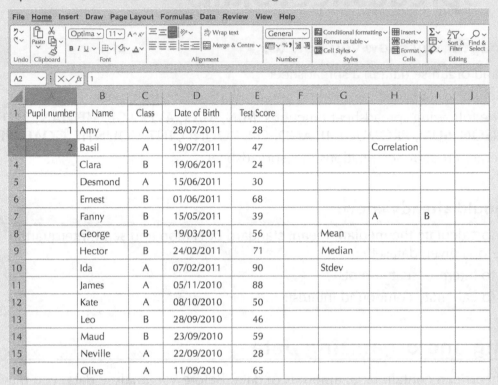

b To sort data you need to highlight the selection of data you wish to sort, then "Sort & Filter", and then "Filter".

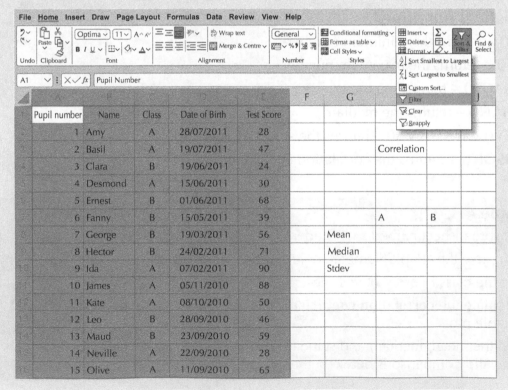

Now you can click the box in the corner of "Date of Birth" cell and select to sort "Newest to Oldest."

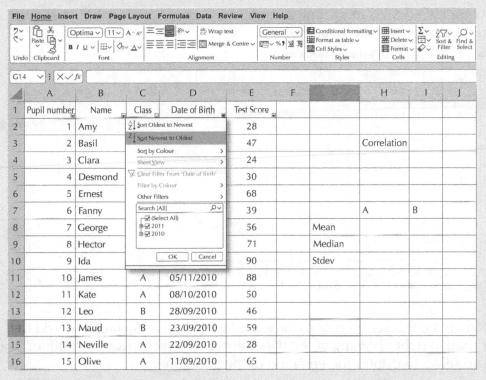

c   Correlation is measured on a scale of −1 to 1; −1 is a perfect, negatively linear relationship, 0 is no correlation, and 1 is a perfect, positive linear relationship. You will cover correlation in detail in Chapter 9. In Excel, you can calculate the correlation coefficient using the PEARSON function.

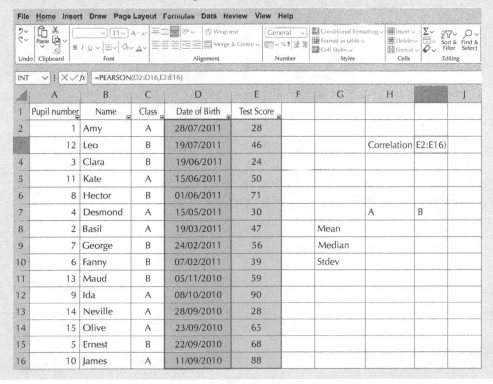

(continued)

The PEARSON function takes two inputs: the first selection of data, and then the second.

Your output is −0.565, suggesting a moderate, negative correlation. A scatter plot can help you see this relationship:

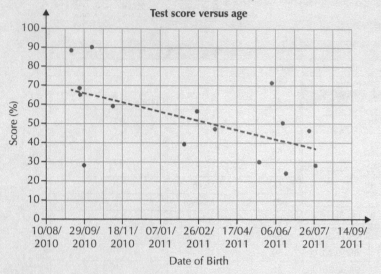

You can create this scatter plot by pressing "Insert" then "Scatter". To add the linear model, right click any of the datapoints and select "Insert Trendline." You can also select to display the equation of the trend line.

d   Now filter data to help make comparisons. Select the box in the corner of the "Class" cell and select "A".

	Pupil number	Name	Class	Date of Birth	Test Score
1	Pupil number	Name	Class	Date of Birth	Test Score
2				28/07/2011	28
3				19/07/2011	46
4				19/06/2011	24
5				15/06/2011	50
6				01/06/2011	71
7				15/05/2011	30
8				19/03/2011	47
9				24/02/2011	56
10				07/02/2011	39
11	13	Maud	B	05/11/2010	59
12	9	Ida	A	08/10/2010	90
13	14	Neville	A	28/09/2010	28
14	15	Olive	A	23/09/2010	65
15	5	Ernest	B	22/09/2010	68
16	10	James	A	11/09/2010	88

Filter menu overlay:
- Sort A to Z
- Sort Z to A
- Sort by Colour
- Sheet View
- Clear Filter From 'Class'
- Filter by Colour
- Test Filters
- Search [All]
- (Select All)
  - ☑ A
  - ☐ B
- OK   Cancel

Now you can select the scores and copy them so that they appear under the "Class A Scores" column. You can repeat this process for the class B scores.

| File | Home | Insert | Draw | Page Layout | Formulas | Data | Review | View | Help |

Undo | Clipboard | Font | Alignment | Number | Styles | Cells | Editing

G18 | fx | 28

	A	B	C	D	E	F	G	H	I	J
1	Pupil number	Name	Class	Date of Birth	Test Score					
2	1	Amy	A	28/07/2011	28					
5	11	Kate	A	15/06/2011	50					
7	4	Desmond	A	15/05/2011	30			A	B	
8	2	Basil	A	19/03/2011	47		Mean			
12	9	Ida	A	18/10/2010	90					
13	14	Neville	A	28/09/2010	28					
14	15	Olive	A	23/09/2010	65					
16	10	James	A	11/09/2010	88					
17							Class A Scores	Class B Scores		
18							28			
19							50			
20							30			
21							47			
22							90			
23							28			
24							65			
25							88			
26										

Using filtering, then copying and pasting, is a simple way to separate data in Excel so that you can compare two different samples of data.

Finding the averages and standard deviation is achieved using the formulae given in the **Statistical Functions** help box.

	A	B
Mean	53.25	51.86
Median	48.50	56.00
Stdev	25.53	16.69

So the final comparison of the scores for class A and class B is: Class A score's have a wider spread (as the standard deviation is greater). It's not clear which class scored higher on average, as class A had a higher mean score, but class B has a higher median. You will consider which average to use in Chapter 7.

Exercise 4A

⬇★ 1 Open the spreadsheet named "4A1".

    a Complete the "Pupil Number" column.

    b Sort the class into alphabetical order by name.

    c Calculate the correlation coefficient between number of hours slept and test score, and create a suitable plot.

    d Find the mean, median, minimum, maximum and sample standard deviation of scores for Mr Jones' and Mr Smith's classes. Make two valid comparisons between the two classes.

    e Make a comparison between how many hours of sleep students get in Mr Jones' class compared with Mr Smith's class.

⬇★ 2 Open the spreadsheet named "4A2". In the column "Day_of_week", the number 1 represents "Monday", 7 represents "Sunday." The spreadsheet contains data about babies born in the USA*.

    a Sort the spreadsheet to show you births of babies born on Friday the 13th. Copy these births into the "Answers" workbook.

> ⚠ You will need to filter two fields: the day of the week and the day of the month.

    b Repeat this for babies born on Friday the 14th and Friday the 15th; copy your results into the relevant columns.

    c Calculate the mean and median for each column. Is there evidence that there are fewer babies born on Friday the 13th?

    d What other questions could you ask about this dataset?

Simple spreadsheet functions

You need to know some simple spreadsheet functions.

> **Simple Excel functions**
>
> The four operations $\times \div + -$ can be used by typing $* / + -$.
>
> **SUM([Range])** – Adds up all the numbers in a list.
>
> **PRODUCT([Range])** – Multiples all the numbers in a list.
>
> **IF([Condition], [Output if true], [Output if false])** – Returns a different output depending on whether the input is true or false
>
> **AND([Condition], [Condition], ...)** – If all conditions are true, it returns true. Otherwise, it returns false.

---

* Data from Five Thirty Eight: https://data.fivethirtyeight.com/ https://github.com/fivethirtyeight/data/tree/master/births.

> **OR([Condition], [Condition], …)** – If any conditions are true, it returns true. Otherwise, it returns false.
>
> **ROUND([Number], [Decimal Place])** – Rounds a number to a given decimal place.
>
> **INT([Number])** – Rounds to the nearest integer.
>
> **COUNTIF([Range],[Condition])** – Count the number of items that meet a condition.

Example 4.2

Open the spreadsheet called "4.2 Money".

Part A

Round the amounts to the nearest penny, and to the nearest pound.

State whether the amount is greater than £5, less than £8, or between the two values.

Count the number of credits and debits.

Find the total of each column.

Part B

Find the sum and product of numbers 1 and 2.

Use **Goal Seek** to fill in the yellow boxes so that the products are: 24, 25, 90, 3.141592654, 1.414213562.

Part C

For each amount, calculate how much money would be in the bank account after 4 years when interest is paid yearly at 3%. Find the total amount of interest earned.

Part A

In cell C3 you use the formula =**ROUND**(B3, 2). This will round the amount to two decimal places. You can use the **INT** function to round to the nearest pound, Cell D3=**INT**(B3).

You can use the IF function to determine whether the amount is greater than £5 or less than £8. Cell E3=**IF**(B3>5, "Yes", "No").

> The B3 > 5 is the condition that Excel will check is true or false. Here you are asking Excel to display "Yes" if true or "No" if false.

Counting the number of credits or debits can be done using **COUNTIF**. Cell I3=**COUNTIF**(A3:A20, "Credit").

> This function counts how many times the word "Credit" appears in all the cells between A3 and A20.

Totals are found using the **SUM()** function.

Part B

For the first three sets of numbers, you can use the **SUM()** and **PRODUCT()** formulae.

(continued)

For the yellow boxes, you can use **Goal Seek.**

First, drag down the formulae for the sum and product.

23			**Part B**	
24				
25	Number 1	Number 2	Sum	Product
27	4	8	=SUM(A26:B26)	=PRODUCT(A26:B26)
27	−4	8	=SUM(A27:B27)	=PRODUCT(A27:B27)
28	5	8	=SUM(A28:B28)	=PRODUCT(A28:B28)
29		8	=SUM(A29:B29)	=PRODUCT(A29:B29)
30		8	=SUM(A30:B30)	=PRODUCT(A30:B30)
31		8	=SUM(A31:B31)	=PRODUCT(A31:B31)
32		8	=SUM(A32:B32)	=PRODUCT(A32:B32)
33		8	=SUM(A33:B33)	=PRODUCT(A33:B33)

To use **Goal Seek**, select the cell you want to meet a particular goal. In your case, this is cell D29 which you want to be 24.

Now click **Data, What-If Analysis, Goal Seek.**

File Home Insert Draw Page Layout Formulas Data Review View Help

Get Data | Refresh All | Queries & Connections, Properties, Edit Links | Stocks Currencies | Sort Filter | Clear, Reapply, Advanced | Text to Columns | What-If Analysis | Forecast Sheet | Outline

Get & Transform Data | Queries & Connections | Data Types | Sort & Filter | Data Tools

Scenario Manager...
Goal Seek...
Data Table...

D29 =PRODUCT(A29:B29)

	A	B	C		E	F	G	
1				**Part A**				
2	Type	Amount	Rounded (Penny)	Rounded (Nearest £)	Greater than £5	Less than £8	Between £5 and £8	Number of Credits:
3	Credit	£4.3929	£4.39	£4.00	No	Yes	FALSE	Number of Debits:
4	Credit	£9.3131	£9.31	£9.00	Yes	No	FALSE	
5	Credit	£5.3164	£5.32	£5.00	Yes	Yes	TRUE	
6	Debit	£5.0087	£5.01	£5.00	Yes	Yes	TRUE	
7	Debit	£9.3513	£9.35	£9.00	No	No	FALSE	
8	Credit	£8.4545	£8.45	£8.00	No	No	FALSE	
9	Debit	£3.0475	£3.05	£3.00	Yes	Yes	FALSE	
10	Credit	£9.6121	£9.61	£9.00	No	No	FALSE	
11	Debit	£4.2582	£4.26	£4.00	Yes	Yes	FALSE	
12	Debit	£9.1796	£9.18	£9.00	No	No	FALSE	
13	Debit	£6.8437	£6.84	£6.00	Yes	Yes	TRUE	
14	Debit	£6.5643	£6.56	£6.00	Yes	Yes	TRUE	
15	Credit	£0.2226	£0.22	£0.00	No	Yes	FALSE	
16	Debit	£9.8831	£9.88	£9.00	Yes	No	FALSE	
17	Credit	£5.3450	£5.35	£5.00	Yes	Yes	TRUE	
18	Debit	£0.4460	£0.45	£0.00	No	Yes	FALSE	
19	Credit	£1.1863	£1.19	£1.00	No	Yes	FALSE	
20	Debit	£6.1441	£6.14	£6.00	Yes	Yes	TRUE	
21	Totals:	£104.5695	£104.56	£98.00		Total True:	6	
22								

	Number 1	Number 2	Sum	Product		Part C		
23								
24	**Part B**					**Part C**		
25	Number 1	Number 2	Sum	Product				
26	4	8	12	32		Interest Rate:	3.00%	
27	−4	8	4	−32		Number of years:	4	
28	5	8	13	40				
29		8	8	8		Amount	Total at end of term	Interest Made
30		8	8	8		£45	£50.65	£5.65
31		8	8	8		£678	£763.09	£85.09
32		8	8	8		£22.21	£25.00	£2.79
33		8	8	8		£57,092	£64,257.55	£7,165.55
34								

In the pop-up box set the value to your desired target, in this case 24. Click the white box labelled "By Changing Cell" and then select the cell you want to change, A29. Click "OK".

19	Credit	£1.1863	£1.19	£1.00	No	Yes
20	Debit	£6.1441	£6.14	£6.00	Ye	
21	Totals:	£104.5695	£104.56	£98.00		
22						
23	**Part B**					Part
24						
25	Number 1	Number 2	Sum	Product		
26	4	8	12	32	Interest Rate:	
27	−4	8	4	−32	Number of years:	
28	5	8	13	40		
29		8	8	0	Amount	
30		8	8	0	£45	
31		8	8	0	£678	
32		8	8	0	£22.21	
33		8	8	0	£57,092	
34						

Goal Seek
Set cell: D29
To value: 24
By changing cell: $A$29
OK   Cancel

You can repeat this process for the remaining cells and their targets.

Part C

Often when creating a spreadsheet it makes sense to set up **key** or **base variables**. These are variables that you want to keep referring to in the spreadsheet. Having key variables allows you to avoid having to type the same numbers repeatedly.

In this example the interest rate and the number of years are the key variables.

(*continued*)

The total at the end of the term is found using the formula:

Amount = Principle amount × (1 + Interest rate(%))^N

where N is the number of periods of time the interest rate is applied. In this case this is 4 years.

You can input that into Excel using the formula: =**ROUND**(F30*(1+$G$26)^$G$27, 2).

> You need to round to two decimal places because your answer is a monetary amount.

Notice that when you refer to key variables, you make an **absolute reference** to them. This is where you type a dollar sign ($) before the letter and number of the cell you refer to. **In Excel, you can do this by pressing F4 after selecting a cell.**

An absolute reference means that when you drag your formula down, the copied formula will still refer to your key variable.

The spreadsheet you have produced can now answer compound interest questions. Try changing the key variables and the amounts and watch the spreadsheet update.

Exercise 4B

The file named "**4B**" can be used to answer all the questions in this exercise.

1   A school is collecting £20 from each student going on a school trip. Complete the spreadsheet to find:

   a   The amount of money outstanding for each student.

   b   The total amount of money collected, and the total amount still outstanding.

   c   The number of students who have not paid their full £20 amounts.

2   Each row contains 5 numbers and their mean, median and range. Complete the table. You will need to use **Goal Seek** on the cells highlighted in yellow. Remember that you will need to have placed a formula in those cells before you complete the **Goal Seek**.

> While there is not a function for the range in Excel, consider how the **MAX** and **MIN** functions could help you.

3   How many Thursdays were there in the year 2003?

★ 4   A bank account offers 3% interest p.a. (per annum). Complete the spreadsheet. Remember to round your answer.

★ 5   The spreadsheet shows a sample of 50 students who are part of their school's sport team. It contains data about which team the student is part of, and whether they are right-handed or not. Find the proportion of students on the football team

> If you want to count how many cells in a row have text in them, use a function called **COUNTA**.

who are right-handed. Compare this with the proportion of right-handed students on the rugby team.

6 A cuboid is measured as having a length of 7 cm, a width of 8 cm and a height of 2 cm. The relative errors of the length, width and height measurements are 3%, 2% and 7% respectively. Estimate the relative error of the volume calculation, then complete the spreadsheet to calculate the relative error. Drag down your formulae to complete the other rows.

Recurrence relations

A recurrence relation is a rule for a sequence which refers to the previous term in the sequence. A simple example would be "Start with the number 1. To find the next term, add 3 to your current term and then double it."

This sequence would be: 1, 8 (1 + 3 = 4, then double), 22 (8 + 3 = 11, then double), 50, 106, 218 …

You write this as: $U_{n+1} = 2(U_n + 3)$.

> Read this as "The next term is equal to the current term add three, then multiply that by 2."

Example 4.3

Open spreadsheet "**4.3 Savings**".

Alexandra deposits £500 into a bank account that pays 0.32% interest per month. She plans to deposit £20 into her bank at the beginning of every month. Complete the spreadsheet to find out how much she earns in interest over a year.

First, fill in the key variables. **Remember** – any time you make reference to the key variables, you will press F4 and make this an **absolute reference.**

	A	B	C
1			
2		Initial Deposit	£500
3		Monthly Interest Rate	0.32%
4		Monthly Deposit	£20
5			

For month zero, you just refer to the initial deposit of £500.

In cell C8, input your interest calculation.

	B	C	
1			
2	Initial Deposit	500	
3	Monthly Interest Rate	0.0032	
4	Monthly Deposit	20	
5			
6	**Month**	**Balance before monthly deposit**	**Balance be**
7	0		=$C$2
8	1	=ROUND(D7*(1+$C$3), 2)	=C8+$C$4

(continued)

This formula takes the previous month's value and multiplies it by the interest rate plus 1. Remember that to increase by 0.32%, you multiply by 1.0032. You round to two decimal places in order to show whole pence. Typically decimal fractions of pence are not shown on information to customers.

D7 is not an absolute reference because it is not a key variable. C3 is an absolute reference.

You can calculate the amount after the monthly deposit by adding the C4. Again, C4 is an absolute reference as it is a key variable.

The total interest is the final balance minus the initial deposit and minus the 12 monthly deposits.

Total interest:	=D19–$C$2–12*$C$4

Exercise 4C

⬇ Open spreadsheet "**4C**". All the questions in this exercise will be completed on the spreadsheet.

1 A sequence is generated by summing the previous two terms.

$U_{n+1} = U_{n+1} + U_n$.

For example: 3, 4, 7 (3 + 4), 11 (7 + 4), 18…

Complete the spreadsheet for the given starting numbers.

2 Complete the column titled "Odd Numbers". The rule for the column titled "Sum of Odd Numbers" is to add the next odd number to previous term. What sequence does this recurrence relation generate?

3 Darcey deposits £1000 into her savings account on the 1 January 2020. Every year, on the same date, she deposits £500. She does this for 6 years. If the annual interest rate is 4%, calculate her balance on the 31 December 2025. Calculate the total interest she has earned in this time.

4 Amara deposits £4000 into her savings account, which pays 0.55% interest per month. Every month Amara withdraws £50. How much money will Amara have in her account after 40 months?

★ 5 Caitlin opens a new bank account. In the first month, she deposits £1. Every month, she will deposit twice as much as the month before. Her savings account pays 0.2% interest monthly. Complete the spreadsheet and find her total balance after one year.

★ 6 A shop is having a closing down sale. Every day that passes, the price of each item for sale is reduced by 10%. Find the cost of a coat that normally sells for £200, after 30 days have passed. Create a scatter plot based on your results and fit a suitable trend line. What model is appropriate to use?

★ 7 A company currently has 197 computers in its offices. The IT manager predicts that 4% of the computers will stop working each year, and will have to be replaced. Every August, 7 new computers will be purchased for the company.

a   Complete the spreadsheet to estimate how many working computers the company will have in August 2033.

b   Comment on the relationship between time and the number of working computers.

c   Extend your table well beyond August 2033 and construct a graph to predict how the number of working computers will change over time.

8   A recurrence relation is defined as $U_{n+1} = 0.3U_n + 20$. Generate the first 20 terms when the first term is 6. Display the results on a scatter plot and fit a suitable trend line. Try changing the starting term. What do you notice?

9   A recurrence relation is defined as $U_{n+1} = -1.1U_n$. Generate the first 50 terms when the first term is 0.002. Display the results on a scatter plot. Try changing the coefficient from −1.1 to −0.9. How does the graph change? Predict what the graph will look like when the coefficient is equal to −1.

- I can sort and organise data in a spreadsheet. ★ Exercise 4A Q2
- I can extract basic statistical information and produce charts from data. ★ Exercise 4A Q2
- I can use the following spreadsheet functions: MIN, MAX, AVERAGE, MEDIAN, STDEV, PEARSON, SUM, PRODUCT, IF, AND, OR, ROUND, INT, COUNTIF, GOAL SEEK. ★ Exercise 4B Q4, 5
- I can use spreadsheets to implement and investigate recurrence relationships. ★ Exercise 4C Q5, 6, 7

5 Venn Diagrams and Tree Diagrams

This chapter will show you how to:

- construct and interpret tree diagrams and Venn diagrams
- derive conditional probabilities from tree and Venn diagrams
- determine the combinations of independent events.

You should already know:

- how to solve simple probability problems.

Probability

In **probability,** an **event** is a possible outcome of the situation you are assigning a probability to. When you flip a coin, the coin landing on heads is one possible **event**. Two events are **independent** if one does not affect the probability of the other.

I flip a coin and it lands on heads. I then flip the coin again, and it lands on tails. Flipping a heads, and then flipping a tails, are two independent events; the first coin flip had no effect on the probability of the result of the second.

Tree diagrams

When you consider the probability of multiple events happening, it is often easier to visualise the different outcomes using a tree diagram. The probability of a series of outcomes can be found by **multiplying** along its branch.

To calculate the probability of one branch occurring **or** another branch occurring, you can **add** the two probabilities.

AND/OR rules

For mutually exclusive **independent events** remember these two rules:

AND – To find the probability of two or more events occurring, **multiply** their probabilities.

OR – To find the probability that one event occurs or another event occurs, **add** their probabilities.

The probability of rolling a fair, 6-sided dice and getting a 1 **or** a 6 is: $\frac{1}{6} + \frac{1}{6}$.

You write this as $P(1 \cup 6) = \frac{2}{6}$.

The probability of rolling a 6, and then rolling the dice again **and** then getting a 1 is $\frac{1}{6} \times \frac{1}{6}$.

You write this as $P(1 \cap 6) = \frac{1}{36}$.

Example 5.1

Sarah is coding a computer game in which a player has two opportunities to open a box and win a prize. The player has a 20% probability of winning a prize from box 1. The player has a 40% probability of winning a prize from box 2. Sarah displays this information in a tree diagram:

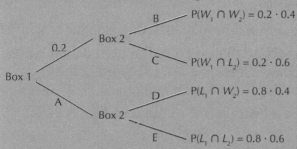

$P(W_1 \cap W_2)$ means: "The probability that the player wins both box 1 **and** box 2."

1   Write down the value of A, B, C, D and E.

2   Find the probability of a player winning a prize from both boxes.

Sarah wants to code her game so that a player has a greater than 50% probability of winning a prize in at least one box. Has she successfully done this?

1   A is the probability of losing with box 1, so 0.8. Notice that the two branches stemming from box 1 must sum to 100%.

   B and D are both 0.4, C and E are 0.6.

2   The probability of winning at both boxes can be found by following the very top branch of this tree diagram. Multiplying along that branch you get $0.2 \times 0.4 = 0.08 = 8\%$.

The probability for the second highest branch (win then lose) is $0.2 \times 0.6 = 0.12 = 12\%$. The probability for the third branch is $0.8 \times 0.4 = 0.32 = 32\%$. This means that the total probability of a player getting at least one win is $8\% + 12\% + 32\% = 52\%$. So Sarah has correctly coded her game.

An alternative way to consider this is that there is only one branch in which the player does not win at least once, and that is the final branch: $P(L_1 \cap L_2) = 0.8 \times 0.6 = 0.48$. So the player fails to win at least once 48% of the time. This means that Sarah has correctly coded her game, as the player wins at least once 52% of the time.

Exercise 5A

1   A company has two machines. The probability that Machine 1 breaks down is 5%. Machine 2 also has a probability of breaking down. This is represented in a tree diagram:

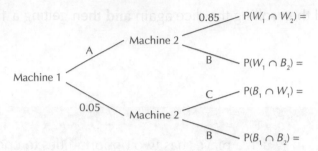

a State the value of *A*, *B* and *C*.

b What is the probability of Machine 2 breaking down?

c Calculate the probability for each branch.

d What is the probability that the company will have at least one working machine?

2 Robert flips a fair coin. He then flips a second biased coin. The biased coin comes up "heads" with a probability of 0.8. Draw a tree diagram to represent the different outcomes. Clearly show the probability of each outcome.

Robert says: "The probability of flipping a head in one of the coin flips is 90%." Comment on what Robert says.

★ 3 A company is producing wind turbines. The production is split into two phases, phase A and phase B. The probability of a delay in phase A is estimated as 5%. The probability of a delay in phase B is estimated as 10%.

a Represent this information in a tree diagram, showing the probability of each branch.

b What is the probability that the wind turbines are produced on time?

4 A deck of cards is made up of 52 cards, including 4 Ace cards. At a charity casino night a game is played in which a player is dealt two cards, without replacement. A player wins a prize if he draws an Ace. The probabilities are represented in the tree diagram:

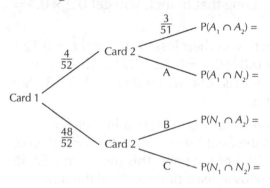

When there is no replacement, this means that the events are not independent. Notice how the numerator and denominator have changed along the top branch for card 2. Why has 4 now become 3?

a Write down the value of *A*, *B* and *C*.

b Calculate the probability of being dealt two Aces.

c Calculate the probability of receiving exactly one Ace.

d Another game at the casino night involves rolling a standard dice, with a player winning a prize if she rolls a 6. In which game is a player more likely to win a prize? Give a reason for your answer.

★ 5 Lauren is going abroad on holiday and is considering purchasing travel insurance. She estimates that there is a 5% probability that she will become sick or injured while on holiday. She estimates that if she does become sick or injured, there is a 30% chance she will need medical treatment while abroad.

    a   Represent this information in a tree diagram.

    b   Find the probability that she requires medical treatment while on holiday.

    c   Lauren estimates the cost of needing medical treatment abroad would be approximately £1363, basing this on the average cost of a medical claim. Lauren can purchase travel insurance, covering medical treatment, for £20. Give two reasons why Lauren might decide to purchase the insurance.

6 There are 4 red balls, 5 blue balls and 3 green balls in a bag. A ball is taken from the bag and not replaced. A second ball is then taken at random.

    a   Represent this in a tree diagram.

    b   Write down the probability that:

        i   Both balls drawn will be red.

        ii   Both balls drawn will be the same colour.

        iii   At least one ball will be green.

    c   Suppose that instead a ball was draw from the bag, then replaced, then a second ball was drawn. Would this make drawing two red balls more likely or less likely than before?

Venn diagrams

Venn diagrams are useful for showing how things fit into different categories, called **sets**. You can see at a glance whether an **element** is a member of more than one set.

Example 5.2

Students were surveyed about which school clubs they regularly took part in. The results are shown in the Venn diagram.

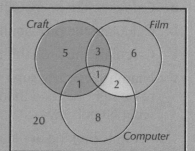

    a   How many students regularly attend the craft club?

    b   How many students regularly attend both the computer and the film club?

    c   How many students were surveyed?

    d   If you choose a student at random, what is the probability that they **do not** attend the film club?

    a   The answer is the total number of students within the "Craft" circle, shaded red in the Venn diagram: $5 + 3 + 1 + 1 = 10$. The answer is not 5, as 5 is the number of students who **only** attend the craft club.

    b   There are 3 students who regularly attend both the film and computer club: 2 students who **only** attend both, and 1 student who attends all 3 clubs. This region is shaded blue in the Venn diagram.

(continued)

> c You can find this by adding up all the numbers; do not forget to include the 20 students who did not attend any clubs. $5 + 3 + 1 + 1 + 2 + 6 + 8 + 20 = 46$.
>
> d There are 12 students who attend the film club, because $6 + 3 + 1 + 2 = 12$. That means that $46 - 12 = 34$ students do not attend. So the probability is $\frac{34}{46} \approx 74\%$.

Exercise 5B

★ 1 Laura carried out some market research, asking a random sample of people if they owned a laptop, desktop or a tablet. She recorded the information in a Venn diagram:

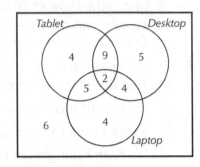

    a How many people said they owned both a tablet and a desktop?

    b How many people said they owned a laptop?

    c How many people did Laura survey?

    d A customer is chosen at random. What is the probability that they own more than one device?

2 Niamh conducts a survey of students to see whether there is a link between being right-handed and liking mathematics. She records her findings in a Venn diagram:

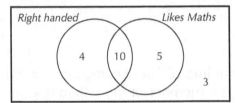

A student is chosen at random.

    a Calculate the probability that the student likes maths.

    b Calculate the probability that the student is right-handed or likes maths.

    c Calculate the probability that the student is right-handed and likes maths.

    d Niamh says: "My survey shows that somebody who likes maths is more likely to be right-handed than left-handed. This suggests there is a link." Comment on what Niamh says.

★ 3 Cameron is looking at data about patients involved in a clinical trial:

	Blood type O negative	
	Yes	**No**
Treatment successful	30	77
Treatment unsuccessful	60	21

    a Construct a Venn diagram to display this information.

    b A patient with blood type O negative is chosen at random. What is the probability that their treatment was successful?

   c   A patient that does not have O negative blood is chosen at random.
         What is the probability that their treatment was successful?

   d   A patient is chosen at random. What is the probability that their treatment was
         successful?

   e   Cameron says: "Patients with blood other than O negative were more likely to
         be successfully treated than those with O negative blood." Comment on what
         Cameron says.

★ **4**   A group of S5 students were asked which apps they had installed on their phones.
         There results were as follows:

   - 58 students had Snapchat installed on their phones.
   - 67 students had Instagram installed.
   - 40 students had TikTok installed.
   - 40 students had both Snapchat and TikTok installed.
   - 45 students had both Instagram and TikTok installed.
   - 35 students had both Snapchat and Instagram installed.
   - 25 students had Snapchat, Instagram and TikTok installed.
   - 5 students had none of these apps installed.

     a   Display this information in a Venn diagram.

     b   A student is chosen at random. Find the probability they have Snapchat
           installed but not TikTok.

   - I can construct tree diagrams and Venn diagrams. ★ Exercise 5A Q3
   - I can interpret tree diagrams and Venn diagrams. ★ Exercise 5B Q1
   - I can derive conditional probabilities from tree and Venn diagrams. ★
     Exercise 5A Q5, Exercise 5B Q4
   - I can determine the combinations of independent events. ★ Exercise 5A Q5,
     Exercise 5B Q3

6 Sourcing Data

This chapter will show you how to:

- identify types of data
- explain the difference between a population and a sample
- identify biases introduced by non-random samples
- explain the influence of outliers on datasets
- compose a research question for your project.

You should already know:

- how to interpret a boxplot
- how to analyse and compare two or more datasets using mean, median, standard deviation and interquartile range.

Gathering data

Types of data

Data can be **categorical** or **numerical**. **Categorical** data are of the form of a label, or category, whilst **numerical** data concern numbers. For example, if you surveyed your class and asked them about their height and eye colour, their height would be **numerical** data and their eye colour would be **categorical**.

Numerical data can be either **continuous** or **discrete**. **Continuous** data can take any value, including non-integer values. For example, 4.5 is a non-integer value, as it lies between two whole numbers. Height is a **continuous** data type, a person's height can be a non-integer like 175.4 cm tall. How many siblings a person has is a **discrete** data type – the answer can only be a whole number. **Discrete** data are measured in integers.

It's important to understand that **continuous** data are always given to a particular level of **precision**. Your height may be 141.2 cm, but if you measured to a greater degree of precision you might have found it to be 141.23 cm.

Example 6.1

Hayley is conducting a medical trial which collects different types of data from patients. Some data are listed in the table:

Gender	Height (cm)	Age (as of last birthday)	Blood type	Peak flow	Times travelled abroad in the last 5 years
M	180	37	O+	503.2	0
F	145	25	O-	466.1	3
M	201	24	AB+	600	1

For each column heading, state which type of data it contains:

- categorical
- discrete numerical
- continuous numerical.

The categorical data columns are gender and blood type. These columns do not contain numbers.

Height and peak flow are continuous data types. Height is still a continuous data type even though it appears to have been rounded to the nearest centimetre.

Age, in this context, is a discrete value. This is because the age is recorded "as of last birthday", meaning that it will only take on integer values.

The number of times the person has travelled abroad is also a discrete value.

Samples and population

A **population** is the entire group you wish to study. A **sample** is the specific group that you get data from. **Populations** can be very large, and so you use **samples** to make **estimations** about the **population**.

If you were interested how S6 students in Scotland planned to vote in the next election, it would be impractical to ask every single S6 student. You could instead ask *some* S6 pupils how they planned to vote and use this as a **model** to help you **estimate** how all S6 pupils might vote.

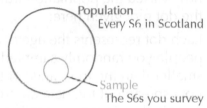

The **population** is every single S6 student in Scotland. The **sample** is the group of students you survey.

Example 6.2

A survey was conducted to find out the voting intentions of S6 students. A random sample of 100 S6 students were asked which political party they preferred. The table shows the results:

Party	Conservative	Green	Labour	Lib Dem	SNP	Other
%	16	21	19	12	29	3

a   What type of data are shown?

b   State what the **population** and **sample** refer to in this study.

c   Which of the following statements is correct?

   A   19% of 17- and 18-year-olds prefer the Labour party.

   B   An estimated 29% of S6 students in Scotland prefer the SNP.

   C   An estimated 12% of S6 students sampled preferred the Lib Dems.

(continued)

a A person's voting intention is an example of **categorical** data. The numbers you can see in the table tell you about the **proportion** of students who favour the different parties.

b The **population** is every S6 student in Scotland. The **sample** is the 100 S6 students who were surveyed.

c Statement A is false – the population is not "17- and 18-year-olds", not every 17-year-old will be an S6 student.

Statement B is correct – as 29% of the randomly sampled students preferred the SNP, you can use this as your **model** to **estimate** what proportion of the population prefer the SNP.

Statement C is false – you know as a fact that precisely 12% of the S6 students sampled preferred the Lib Dems. **This is not an estimate**.

Estimating the standard deviation of a population is not as simple as calculating the standard deviation of your sample. This is because a sample will often underestimate how **varied** the population truly is. Consider the dot plot given here:

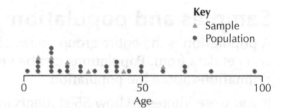

Key
▲ Sample
● Population

Each dot represents the age of a person in the population. The green dots are the people you randomly sampled. Notice that the **range** of the sample would be much smaller than the true range of the population. The standard deviation, a measure of how spread out the data are from the mean average, would also be smaller in the sample than in the population.

For this reason you use a slightly different formula for standard deviation depending on whether you are **estimating** a **population's** standard deviation based on a sample, or whether you are **calculating** the standard deviation of an entire population. In Excel, use **STDEV.S** when you have a **sample**, use **STDEV.P** when your dataset is the entire **population**.

Exercise 6A

★ 1 Zoe takes a random sample of books from her school library. Part of her dataset is shown in the table.

Gender of author	Genre	Gunning Fog Index	Average number of letters per word
M	Horror	8.8	4.3
F	Adventure	11.2	5.2
F	Fantasy	14.7	4.5

Zoe finds the mean Gunning Fog Index for the dataset is 9.78.

a State the data type of each column.

b State what the population and sample are for this dataset.

c Zoe says: "I estimate that the average Gunning Fog Index for school libraries is around 9.8." Comment on what Zoe says.

2 Joshua is reviewing his company's recent data. He sees that in the last month his company manufactured 132 fridges and had sales of £456 214. Describe these kinds of data.

3 Write down an example of data that would be:

a numerical

b discrete

c continuous

d categorical.

★ 4 Grace is investigating the sleeping habits of teenagers in Scotland. As part of the investigation, Grace surveys 100 students at her school. The results of her investigation will be published on her data blog. State what "population" and "sample" refer to in this investigation.

5 Fern is interested in finding out the standard deviation for the heights of students in her class. She measures every student and types the data into a spreadsheet. Which formula would she use to calculate the standard deviation?

6 Ramsey asks a random sample of students in his school about how many books they have read this year.

a What type of data would he obtain?

b State the population and sample in this context.

c Which formula should Ramsey use to calculate the standard deviation for books read by students in his school?

7 A environmental charity are conducting a survey asking the British public what they believe to be the most important factors for sustainability. The charity took a random sample of 3264 British people. Part of their results are shown in the table.

Factor	Frequency
Optimise current use of fossil fuels	593
Eliminate waste	453
Recycling	775
Reducing pollution	800
Carbon capture technology	643

a State the data type that best represents the factors the public see as important for sustainability.

b A charity worker says: "The majority of Scots believe reducing pollution is the most important factor for sustainability." Give two reasons why this view cannot be supported.

Bias

You take **samples** to make estimates about the **population**. For these estimates to be accurate, the sample must be *representative* of the population. You achieve this by taking a **random** sample. A sample is **biased** when some parts of the population are more likely to be included in the sample than others. Conducting a survey at 2pm on a weekday would introduce bias into a sample as people who go to school, or work in a traditional nine-to-five job, would be much less likely to be included in the sample.

A **simple random sample** is taken by assigning every member of the population a number, and then choosing the sample by picking numbers at random.

Example 6.3

Caleb is conducting an opinion poll to gauge how people would vote in a referendum on Scottish independence. He wishes to investigate the views of people who are eligible to vote. One afternoon Caleb goes to his town centre and stops and surveys 50 people.

a   Explain why this method might introduce bias to the sample.

Caleb decides to create his own online survey. He writes the following question: "Given the current turmoil in Westminster, do you favour Scottish independence or remaining in the UK?"

b   Explain why this question is problematic and rewrite it.

a   His own town may not be representative of all of Scotland. He is more likely to survey people who live in a town rather than more rural voters or city dwellers. The time and day would bias the sample; for example, Monday at 11:00am would mean office workers could not be sampled.

b   Caleb's question is a leading question that may encourage more people to respond in favour of Scottish independence. A better question would be: "If a referendum were held tomorrow on Scottish independence, how would you vote? Yes or no to Scottish independence?"

Exercise 6B

1   A company surveys its customers by asking them to fill out an online survey. How might this method introduce bias into the sample?

★ 2   A headteacher wants to know what proportion of her students walk to school. She decides to take a sample by asking the first 30 students who enter the school gates. Explain why this method could introduce bias into the sample, and briefly suggest a better method.

3   Elsie wants to see what the voting preferences of the students in her school are. She makes the following post on social media:

Which party would you vote for?

- SNP
- Labour
- Green
- Liberal Democrats

Give three reasons why Elsie should not consider the result of this poll as a good estimate for the preferences of the students in her school.

4   Lucas listens to a British radio station that asks its listeners to go online and vote in its daily poll, which today is about whether income tax should be lowered. At the end of the radio show, the presenter shares the result of poll.

Write down three groups of people in the UK population who would not be present in the sample.

5   A charity that campaigns against climate change wishes to survey residents of Glasgow to find out whether they believe climate change is real or not. The charity set up a stall on a busy shopping street and asks shoppers to fill out a paper survey asking: "Do you believe that climate change is real? Yes or No."

The charity later publish a statement saying: "97% of Glaswegians believe climate change is real."

Give two reasons why this statement is unlikely to be valid.

The influence of outliers

Outliers are data points that significantly differ from the rest of the data in a dataset. Outliers can strongly influence your data analysis. If an outlier has been caused by an error, for example, when measuring something or typing the data into a spreadsheet, it is acceptable to remove it. It is not acceptable to ignore an outlier just because it does not fit a trend.

Outliers have a large effect on the **mean** and **standard deviation**. The **median** and **interquartile range** are less affected by outliers.

Outliers will also affect the **correlation coefficient**.

Correlation coefficient

The correlation coefficient is a measure, from −1 to 1, of how correlated two variables are. The closer the value is to −1, the stronger the negative correlation. The closer the value is to 1, the stronger the positive correlation. The closer the value is to 0, the weaker the correlation.

A correlation of around −0.9 shows a clear, negative correlation.

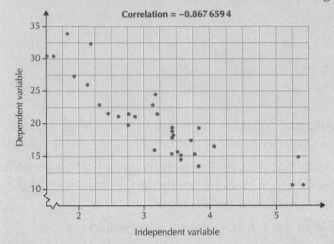

A correlation around −0.1 shows no apparent correlation.

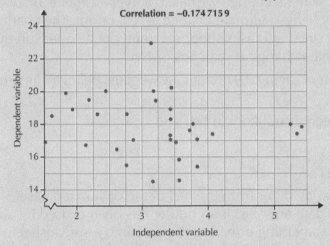

A correlation close to 1 shows a strong, positive correlation.

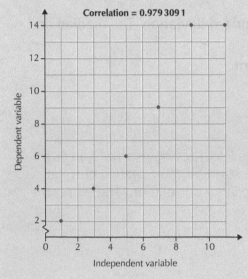

Example 6.4

Noah is investigating whether there is a correlation between the number of bedrooms a residential property has and its selling price. He produces a scatter plot.

Explain which two data points can be removed as input errors. How will the correlation coefficient change when these points are removed?

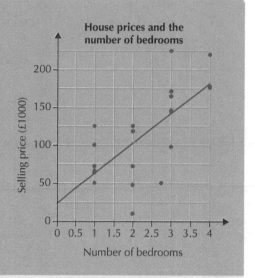

The two outliers to be removed are the points around (2.0, 10) and (2.75, 50). For the first point the selling price is small enough to not be plausible. The second point is for "2.75" rooms, which should be a integer value.

As both these points make the scatter plot less correlated, when Noah removes them the correlation coefficient will increase.

Exercise 6C

★ 1 Lucy is investigating the birth weights of newborn babies. She produces the following histogram:

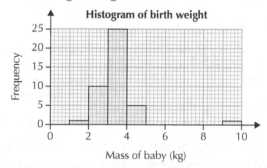

Lucy notices that her dataset has one entry for a 10 kg newborn, and decides to remove that entry.

a Is it acceptable for Lucy to remove this outlier?

b What effect will removing this outlier have on the mean?

c Which measure of spread will be more affected by the removal of the outlier: the standard deviation or the interquartile range?

2   Arthur is analysing data about the vocabulary of children. His data can be found in the spreadsheet file **6B**. Here is some of his analysis.

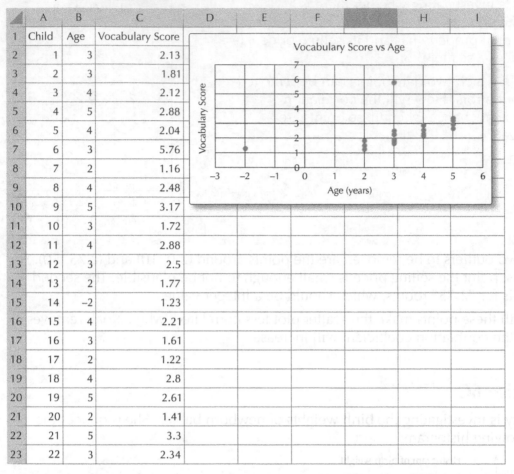

	A	B	C	D	E	F		H	I
1	Child	Age	Vocabulary Score						
2	1	3	2.13						
3	2	3	1.81						
4	3	4	2.12						
5	4	5	2.88						
6	5	4	2.04						
7	6	3	5.76						
8	7	2	1.16						
9	8	4	2.48						
10	9	5	3.17						
11	10	3	1.72						
12	11	4	2.88						
13	12	3	2.5						
14	13	2	1.77						
15	14	−2	1.23						
16	15	4	2.21						
17	16	3	1.61						
18	17	2	1.22						
19	18	4	2.8						
20	19	5	2.61						
21	20	2	1.41						
22	21	5	3.3						
23	22	3	2.34						

Chart title: Vocabulary Score vs Age. Y-axis: Vocabulary Score. X-axis: Age (years).

a   Identify which two children could be classified as outliers.

b   Can either of these outliers be removed?

c   What effect would removing these outliers have on the correlation coefficient?

> ⚠ Remember that the correlation coefficient is a measure, from −1 to 1, of how correlated two variables are.

3   Arabella produces a histogram of the number of customers who have visited her different business locations. Here is part of her spreadsheet, which can be found in spreadsheet **6B**.

	A	B	C	D	E	F	G	H	I
1	Shop	Number of customers				Mean	1514.32		
2	A	1627				Median	1598.5		
3	B	1163				Range	958		
4	C	1843				STDEV	295.7789		
	D	1764							
6	E	1514.32							
7	F	1868							
8	G	1576							
9	H	1404							
10	I	1750							
11	J	1382							
12	K	1086							
13	L	1674							
14	M	1673							
15	N	1283							
16	O	1123							
17	P	1245							
18	Q	1254							
19	R	1827							
20	S	1999							
21	T	1050							
22	U	1204							
23	V	1041							
24	W	1721							
25	X	1621							
26	Y	1702							
27	Z	1978							

Histogram of Customer Numbers

Frequency (y-axis: 0 to 4.5)

Customer Numbers (x-axis): [1041,1141], [1141,1241], [1241,1341], [1341,1441], [1441,1541], [1541,1641], [1641,1741], [1741,1841], [1841,1941], [1941,2041]

a Which shop's data should be removed?

b How would removing the outlier effect the mean, median, range and standard deviation?

4 Emma visits several burger restaurants and makes a note of the cost of a cheeseburger at each location. She displays her results in a stem and leaf diagram.

Stem and leaf diagram of cheeseburger prices

```
  3 | 4 5 6        Key: 3 | 4 = £3.40
  4 | 1 3 3 7
  5 | 0 1 1
  6 | 2 2
400 | 0
```

a Find the median cost of a cheeseburger in this sample.

b Which burger price can be removed?

c Find the median after the outlier has been removed.

★ 5   Rebecca is looking at data that shows how customer feedback scores have changed over time (see Spreadsheet 6B).

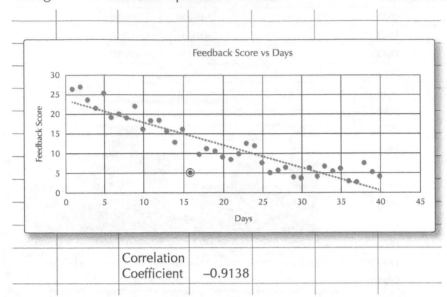

Correlation Coefficient	−0.9138

She decided to remove the point circled in red as an outlier.

a   How will the correlation coefficient change when the outlier is removed?

b   Comment on Rebecca's decision to remove the outlier.

Starting your project

In your project you will choose a topic to investigate. You will have to find a dataset to analyse. Your teacher will check that your topic and dataset are appropriate before you write your report. Think carefully about a topic that you find interesting or know something about.

Example 6.5

Harry is interested in finding out whether people who work in education are more likely to work longer hours than people in other industries. Harry searches online and downloads a dataset from the National Records of Scotland. You can see the downloaded file in the spreadsheet named **6.5***.

Harry considers the following questions:

- What is the dataset about?
- What do the column headings mean?
- How robust are the data? Who was the source of the data and by what method did they gather the data? Is this credible?
- Is the dataset a sample or a population?

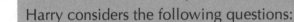

* National Records of Scotland licensed under the Open Government Licence. Downloaded from https://www.scotlandscensus.gov.uk/documents/microdata-teaching-file-and-user-guide/

- What questions could you ask about the dataset?
- What type of data does your question relate to?

Take your time to look at the dataset and the "Column Meanings" tab. It is easy to be overwhelmed at first, so start by thinking about what the columns mean.

The dataset shows lots of information about a sample of people in Scotland, including their marital status, the industry they work in, whether they work full time and whether they work more than 49 hours a week.

The column headings can be understood by looking at the "Column Meanings" tab.

The data are very robust. Their source is the National Records of Scotland, and they gathered this data by randomly sampling 1% of Scotland's population and speaking with them.

Examples of questions include: "Are people born outside of the UK more likely to work in the construction industry?", "Are Buddhists more likely to be in very good health compared with Christians?", "Are women more likely to be employed part time than men?". Harry's intended question about the hours worked in Education is a possible question.

Harry is interested in **categorical data**. Those categories are "Full-time 31–48 hours worked" and "Full-time 49 or more hours worked."

For Harry to make a comparison, he will need to **filter** the dataset and copy items onto a new worksheet. He filters education workers using the "industry" column.

He also filters to only consider full-time workers.

(continued)

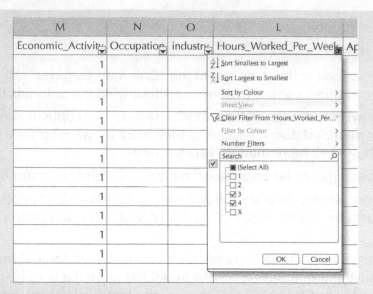

He copies this over into the "Harry's Working" tab. He repeats the process, this time filtering for full-time workers in every other industry.

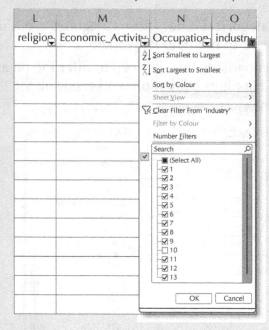

What other questions could you ask about this dataset? How would you need to sort or filter the spreadsheet to answer them?

Think of a topic that interests you, and search online for a dataset about it. Make sure you choose a dataset which is a sample, not an entire population.

Answer the following questions:

- What is the dataset about?
- What do the column headings mean?
- How robust is the dataset? Who was the source of the data, and by what method did they gather the data? Is this credible?
- What questions could you ask about the dataset?
- What type of data does your question relate to?

Ask your teacher to check your research question and dataset. You can use your answers to the listed questions to help you write your introduction.

- I can identify types of data: categorical, continuous numerical, discrete numerical. ★ Exercise 6A Q1
- I can explain the difference between a population and a sample. ★ Exercise 6A Q4
- I can identify biases introduced by non-random samples and find alternatives to avoid bias. ★ Exercise 6B Q2
- I can explain the influence of outliers on datasets. ★ Exercise 6C Q3, 5
- **I have created a research question for my project.**

7 Forming a Subjective Impression

This chapter will show you how to:

- use RStudio to produce statistical diagrams
- interpret statistical diagrams
- determine, using a histogram, whether data are normally distributed or skewed
- interpret measures of spread and dispersion
- form a subjective impression of your project's dataset.

You should already know:

- the meaning of median, mean, interquartile range and standard deviation
- how to construct simple statistical diagrams by hand.

RStudio

When you first look at a dataset, it is useful to form a **subjective impression**. What does the dataset appear to show you? You can make a subjective impression by producing **statistical diagrams** and calculating the **measures of spread and dispersion** for your sample. R is statistical software that can be used to help you make sense of data. In this book we will use RStudio, which utilises R.

Importing data

When you first open RStudio, you need to open a new R script.

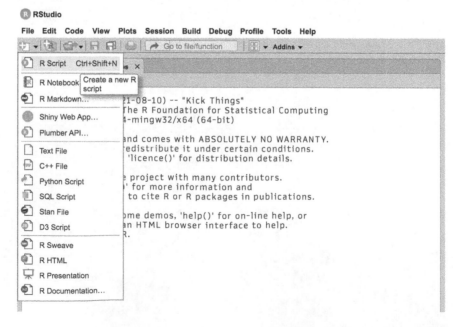

You also need to import your data into RStudio. First, set the working directory to the location where your spreadsheet is saved.

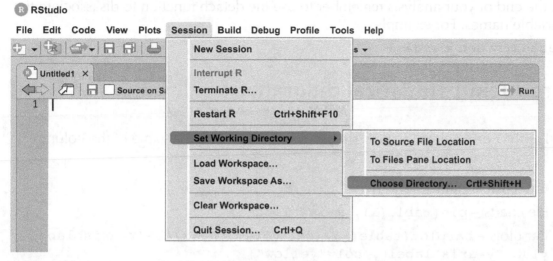

Make sure that your spreadsheet is saved in the "csv" file type.

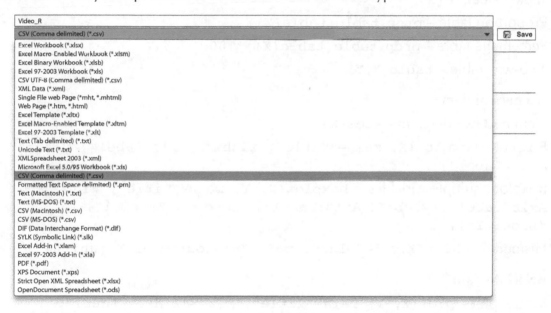

The correct filetype is "CSV (Comma delimited) (*.csv)". Avoid using any other filetype to prevent problems.

Next, you import data into R using the command:

Give your data a suitable name.

```
video_data <- read.csv("Video_R.csv")
```

Then type:

This must be the exact name of your csv file.

```
attach(video_data)
```

To run any code in RStudio, highlight the code you wish to run, then press control and enter at the same time.

What you attach must match the name you gave your data above. In this example, you named your dataset "video_data".

To check that you have correctly imported your data, type: `names(video_data)`. This will display all the names of the columns from your spreadsheet.

At the end of your analysis remember to use the detach function to disassociate the variable names. For example:

```
detach(video_data)
```

Producing statistical diagrams

For the code below, X and Y should be replaced with the name of the column heading you are analysing.

Categorical data

Pie charts – `pie(table(X), main="Title")`

Bar plots – `barplot(table(X), main="title", xlab="x-axis label", ylab="y-axis label", col="yellow")`

Tables – `table(X)`

Proportion table – `prop.table(table(X))`

Percentage table – `prop.table(table(X))*100`

Two-way table – `table(X,Y)`

Numerical data

Stem and leaf diagram – `stem(X)`

Boxplot – `boxplot(X, main="Title", xlab="X Axis Label", ylab="Y Axis Label")`

Boxplot (multiple variables) – `boxplot(X, Y, main="Title", xlab="X Axis Label", ylab="Y Axis Label", names=c("Label 1", "Label 2"))`

Histogram – `hist(X,col="blue", main="Histogram of X (units)")`

Bivariate data

Scatter plot – `plot(X, Y, main="Title", xlab="X Axis Label", ylab="Y Axis Label", pch=19)`

You do not need to memorise this information. In the exam you will be supplied with R code which you will be able to copy and paste into RStudio. When completing your project you can look up information. The **Appendix** has more detailed examples of statistical diagrams for you to use when creating your project.

Example 7.1

Import the file "Video_R.csv" into RStudio.

Participants were asked to watch 4 medical videos, and then were asked to score each video.

a Produce a table showing the gender of the participants.

b Represent the gender of the participants in a suitable diagram.

c Produce a two-way table showing the gender of the participants and their favourite video.

d Produce boxplots for the Scores for videos A, B, C and D.

e Make a comparison between the 4 videos.

f Produce a boxplot comparing how male and female participants scored video C.

The R code for this example can be found in the file 71.R.

a This can be created using:

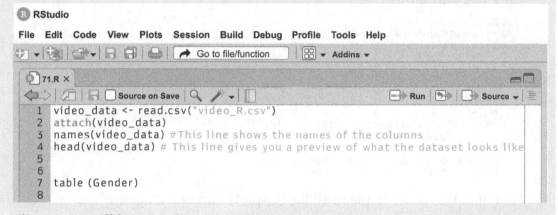

The output will be:

```
> table(Gender)
Gender
Female    Male
    13       7
```

If you would prefer to see the output as a proportion or a percentage, use:

```
 8  prop.table(table(Gender))
 9  prop.table(table(Gender))*100
```

```
> prop.table(table(Gender))
Gender
Female    Male
  0.65    0.35
> prop.table(table(Gender))100
Gender
Female    Male
    65      35
```

From this you can see that 65% of participants were female.

b Gender is a form of **categorical** data, so either a pie chart or a bar chart would be suitable.

```
12  #Pie chart and Barplot
13  pie(table(Gender), main="Pie Chart of Gender")
14  barplot(table(Gender), main="Barplot of Gender", xlab="Gender", col="yellow")
15
```

(continued)

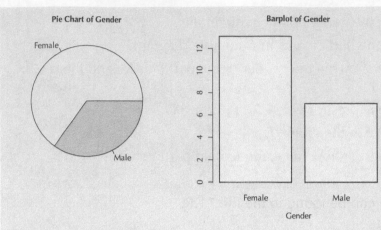

c For a two way table:

```
16  #Two-way table
17  table(Gender, Favourite.Video)
18  prop.table(table(Gender, Favourite.Video))
```

```
> table(Gender, Favourite.Video)
        Favourite.Video
Gender   A  B  C  D
 Female  4  2  3  4
 Male    1  3  2  1
> prop.table(table(Gender, Favourite.Video))
        Favourite.Video
Gender     A     B     C     D
 Female  0.20  0.10  0.15  0.20
 Male    0.05  0.15  0.10  0.05
```

d Here it makes sense to place all the boxplots into one diagram:

```
21  #boxplot
22  boxplot(Score.A, Score.B, Score.C, Score.D,
23          main="Boxplots of scores",
24          xlab="Videos",
25          ylab="Score",
26          names=c("A", "B", "C", "D"),
27          col="Cadetblue2")
```

R has lots of colours. To see what options you can choose, run the code "colors()".

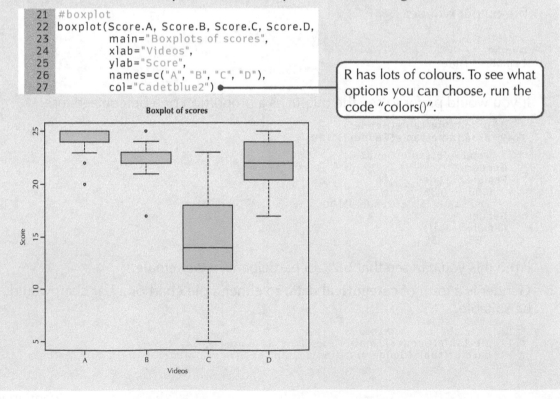

e   Some valid comparisons include: Video A had the highest average (median) score of all 4 videos. The scores given to video C were much more varied than then other 3 videos. Videos A and B had outliers.

f   R can filter in this way when you use a tilde (~):

```
34  boxplot(Score.C ~ Gender,
35          main="Boxplots of Video C's Score by Gender",
36          xlab="Gender",
37          ylab="Score",
38          col="saddlebrown")
```

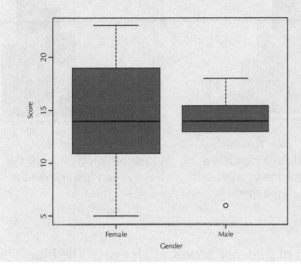

Boxplot of Video C's Score by Gender

Exercise 7A

Open a new R Script and import the dataset called "7A.csv". This dataset has information about people who followed one of three diets. It contains information about their gender, age, height, which diet they followed (1, 2 or 3), their pre-diet weight, their weight after six weeks on the diet and the total amount of weight they lost.

1   Produce a table showing the gender of the participants. What percentage of participants were female?

2   Produce a two-way table showing gender and diet.

★ 3   Represent the gender of participants on a suitable chart.

4   Represent the number of participants on each diet on a suitable chart.

5   Produce a boxplot showing the weights of participants before and after their diet.

6   Produce a boxplot showing how much weight was lost by each diet. **Remember to use a tilde**.

7   The researchers want to know which diet is the most successful. Use your answer to part 6 to give your subjective impression.

8   Produce a scatterplot for the participants' pre-diet weight and weight after six weeks on the diet.

Histograms

Data are **normal** if they are distributed in a **symmetrical** way and **skewed** if the distribution of data is **not symmetrical**.

You can use a **histogram** to determine how data are distributed.

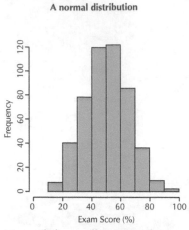

A normal distribution

Normal data will appear close to symmetrical, with the bulk of the data in the centre. The median and the mean of the dataset will be approximately equal.

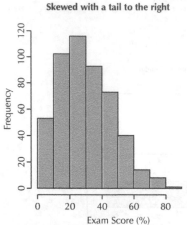

Skewed with a tail to the right

When the distribution has a right tail, the mean will be higher than the median.

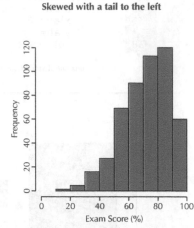

Skewed with a tail to the left

A distribution with a left tail has a mean lower than the median.

The histograms show you the distribution of students' scores in an exam. The blue histogram shows you that around 90 students scored between 60% and 70%. The orange histogram shows you that around 100 students scored between 10% and 20%.

When working with **normal (symmetrical)** data, you should use the **mean** and **standard deviation** as your measures of **location** and **dispersion**.

When working with **skewed** data, you should use the **median** and **interquartile range** as your measures of **location** and **dispersion**. This is because the median and interquartile range are more robust to outliers.

Robust to outliers

Consider the following sample dataset:

1,3,5,7,9

This dataset has a mean of 5, and it also has a median of 5. The interquartile range is 4, and the standard deviation is 3.16.

Consider what would happen if you changed the 9 to 100. How would the mean and median change?

1,3,5,7,100

The median remains 5, but the mean is now 23.2. The interquartile range remains 4, but the standard deviation is now 43.

The median and interquartile range are more robust to outliers. The mean and standard deviation are more influenced by outliers.

Example 7.2

Figure 7.17 is a histogram for Kieran's dataset:

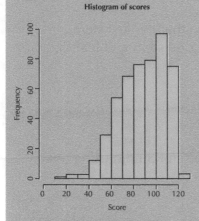

Histogram of scores

a Describe the distribution of Kieran's data.

b State the correct measure of location and dispersion.

c Which of the following stem and leaf diagrams is the best match for Kieran's data?

```
The decimal point is 1 digit(s) to the right of the |

   2 | 3
   2 |
   3 | 4
   3 | 5778
   4 | 23
   4 | 56999
   5 | 01222223333344
   5 | 55667778899
   6 | 000000111222223333334444
   6 | 5555555556666666777788889999
   7 | 00000000011111223334444444
   7 | 5555555555566666667777788888999999
   8 | 0000000011111112222222333333333333344444
   8 | 55555556666666777888888889999999999999
   9 | 0000000000000000011111122222222222223333333334444444444444
   9 | 5555555556666666666777777777778888888888999999999999
  10 | 00000000000001111112222222222223333333344444444444
  10 | 55555555566666666667777777778888888999999999
  11 | 000000000111111122222222333333333444444
  11 | 5555666666666677777888888899
  12 | 0000111
```

OR

```
The decimal point is 1 digit(s) to the right of the |

   0 | 223334444
   0 | 555556666666777788888889999999999
   1 | 0000000000011111111222222222222333333444444
   1 | 555555555666666666666677777777788888888888899999999999999
   2 | 00000011111111112222222222222333333333333334444444444444
   2 | 55555555666666666677777777788888888888888899999999999999
   3 | 0000000001111111222222333333333333344444444444
   3 | 555555555566677777777777777778899999999999
   4 | 0000000011122222233333333334444
   4 | 555555555556666666666677777777899999999
   5 | 00000001111222223333333334444444444
   5 | 555555556667777778999
   6 | 000001111223333334
   6 | 667777788999
   7 | 00011334
   7 | 55778
   8 | 123
   8 | 55899
   9 | 002
```

> **a** Kieran's data are skewed, with a left tail.
>
> **b** The correct measure of location is median. The correct measure of spread is the interquartile range.
>
> **c** Turn this book 90 degrees anti-clockwise. You can see that the first stem and leaf diagram matches the left tail distribution. The second stem and leaf diagram is of skewed data with a right tail.

Measures of spread and dispersion

You can find different measures of spread and dispersion in RStudio using this code:

```
mean(X)
median(X)
IQR(X)
sd(X)
summary(X)
```

Exercise 7B

⬇ Answers can be found in R file 7B.R.

1 Comment on the distribution of the data and state the correct measure of location and dispersion in this histogram.

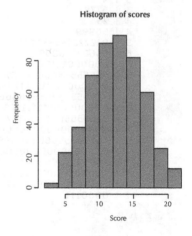

2 Comment on the distribution of the data and state the correct measure of location and dispersion in this histogram.

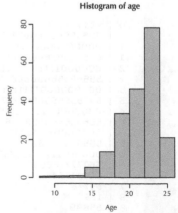

3   Import the file "7A.csv". This is the dataset you used previously in Exercise 7A.

   a   Produce a histogram for the heights of the participants in the study.

   b   Produce a histogram for the ages of the participants in the study.

Remember to include a title for your histogram.

   c   Find the correct measure of location and dispersion for height and age for this study.

4   Import the file "7BQ4.csv" into RStudio. The data show the scores received for two different groups of people.

   a   Produce a boxplot for group A and group B.

   b   Make two valid comparisons between these two groups.

   c   Produce a histogram for each group.

   d   Calculate the correct measure of location and dispersion for each group.

For Questions 5-8, import the dataset called "Ice_cream.csv". This dataset contains information about the gender of the students participating in a study, their favourite flavour of ice-cream, as well as their scores in a video game and puzzle game.

5   Using the Ice_cream dataset:

   a   Create a table showing the number of male and female students involved in this study.

   b   Create a table showing the percentage of students who chose each flavour of ice-cream.

   c   Illustrate the favourite flavours of ice-cream in a suitable diagram.

   d   What was the modal favourite flavour of ice-cream?

   e   Produce a two-way table of gender and favourite ice-cream.

   f   What was the modal favourite flavour for males and females?

   g   Form a subjective impression: Do males and females have different tastes in ice-cream?

6   Using the Ice_cream dataset:

   a   Produce a stem and leaf diagram for the video scores.

   b   How many students got a video score of 42?

   c   What was the modal video score?

   d   By producing a histogram, or otherwise, state whether the video score data are normally distributed.

   e   Calculate the appropriate measure of location and dispersion for the video score.

7   Using the Ice_cream dataset:

   a   Illustrate the students' puzzle scores in a suitable diagram.

   b   Produce a histogram for the puzzle score.

    c Produce a boxplot comparing the male puzzle scores to female puzzle scores.

    d Find the correct measures of location and dispersion for the puzzle scores.

    e Make two valid comparisons between the puzzle scores and the video scores.

    f Form a subjective impression: did students perform better at the puzzle or the video?

8 Using the Ice_cream dataset:

    a Produce a scatterplot comparing video scores with puzzle scores.

    b Form a subjective impression: is there a linear relationship between these variables?

For Question 9 import the dataset named "Cholesterol_R.csv".

Researchers measured the cholesterol (in mmol l^{-1}) in participants before they started eating a particular margarine. They also measured the cholesterol after 4 weeks and after 8 weeks. Participants were given one of two possible margarines to use in their diet: margarine A or margarine B.

★ **9** **a** Produce a histogram for the "before" cholesterol measure.

    b Give the correct measure of location and dispersion for the "before" cholesterol measure.

    c Find the correct measure of location and dispersion for the cholesterol measure after 8 weeks.

    d Display the "before" cholesterol measure and "after 8 weeks" cholesterol measure in a suitable diagram. Make two valid comparisons between them.

    e Form a subjective impression: Did the diet reduce the measure of cholesterol?

    f Produce a table showing the proportion of participants who were given margarine A compared to margarine B. Display this information using a suitable diagram.

    g Produce a scatter plot for "before" and "after 8 weeks". Comment on whether there seems to be a linear relationship.

Forming a subjective impression for your project

Example 7.3

In your project you will have found a dataset and decided upon a question to investigate. The first step of this investigation will be to form your subjective impression. You will need to produce graphical displays and comment upon them, generate descriptive statistics and consider the normality of any numerical data.

Rebecca wants to form a subjective impression. Her dataset is a random sample of salaries of people in her town. Her research question is whether there is a gender difference in pay.

She is forming her subjective impression.

Her data type is numerical (continuous), and so the appropriate graphical display would be a boxplot.

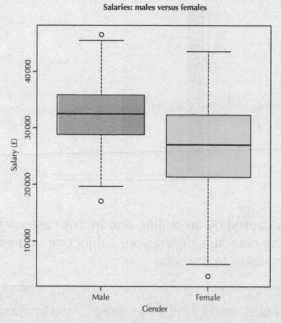

Salaries: males versus females

Her data seem to show that males earned higher salaries on average than females, and that the salaries of females were more varied.

Rebecca wants to know which measure of location and dispersion to use and so she produces histograms.

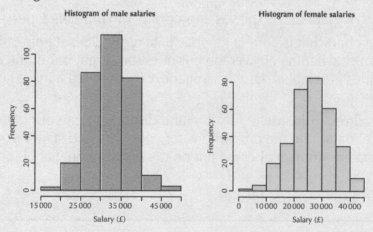

As both histograms are approximately symmetrical, Rebecca decides to use the mean and standard deviation.

Gender	Mean	Standard deviation
M	£32 251	£4913
F	£26 317	£7557

Rebecca forms the subjective impression that women have lower salaries than men, on average, in her town.

Mistakes can be useful. Imagine if Rebecca's histogram had looked like this:

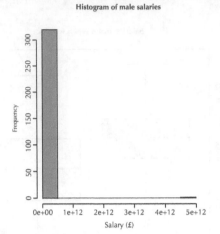

Histogram of male salaries

A histogram like this is caused by an outlier, and in this case such a large outlier that it must be an error in the data. It is during your subjective impression that you will likely encounter any mistakes in the data.

For your project you will have found a dataset you are interested in using and have a research question you are exploring.

The first thing you want to do is form a subjective impression about your data in relation to your question. Consider the following questions:

1 What type of data do you have?

2 What type of graphical display is appropriate to your data type?

3 Produce a chart that is correct for your data type. Does it help you form an impression of what the answer to your question might be? If not, say why not. Include titles and labels. Use the **Appendix** if you need a more complex diagram.

4 State the correct measure of location and dispersion for your data. For categorical data, find the proportions for your data and state the sample size. If your data are numerical comment on the distribution. Label these statistics clearly.

5 State your subjective impression.

- I can use RStudio to produce statistical diagrams. ★ Exercise 7A Q3
- I can interpret statistical diagrams, including stem and leaf, box plots, histograms, pie charts, and bar charts. ★ Exercise 7B Q4, 8
- I can determine, using a histogram, whether data are normally distributed or skewed. ★ Exercise 7B Q2, 8
- I can interpret measures of spread and dispersion. ★ Exercise 7B Q8
- **I have formed a subjective impression of my project's dataset.**

8 Hypothesis Tests, Confidence Intervals and Errors

This chapter will show you how to:

- interpret the results of a hypothesis test
- understand the difference between a type I and type II error
- interpret confidence intervals
- discuss confounding variables.

You should already know:

- the difference between a population and a sample
- how to convert a decimal to a percentage.

Hypothesis testing

Hypothesis testing is used to see whether *chance* is a plausible explanation for the data you see. Suppose you conduct a random sample of men and women and find that 12% of men were left-handed, whereas only 8% of women were left-handed. Is that difference between men and women enough to conclude there really is a significant difference between the sexes?

Using a random sample is a good way of avoiding bias and ensuring that your sample is representative of the population you are sampling. But it is still possible that, by chance, your results are not representative of the whole population. This is called a **sampling error**. How do you know there really is a difference between men and women in the population. What if your random sample just happened to select a few extra left-handed men?

When conducting research you will form a hypothesis you wish to test. For example, "The proportion of men who are left-handed is different to the proportion of women who are left-handed." The hypothesis you wish to test is called the **alternative hypothesis.**

The **null hypothesis** is the opposite of the alternative hypothesis. For example, "There is no difference in the proportion of men and women who are left-handed."

Hypothesis testing will give you a probability, which you refer to as a "***p*-value.**" This is the probability of your sample being as extreme as it is if the null hypothesis was true. To understand this, imagine that your null hypothesis is true: there really is no difference in the proportion of men and women who are left-handed. What is the probability of sampling the population and getting a result as extreme as you did?

You conduct a hypothesis test and get a p-value of 23%. This means, even if there is **no difference** between left-handedness in men and women, there is still a 23% probability of taking a sample that finds a difference as extreme as the one you found (12% of men but only 8% of women). A value of 23% is a high percentage, and so it would not be sensible to conclude that there is actually a difference between men and women. In this case, you **fail to reject** the null hypothesis.

Consider a different example where you take a sample of men and women and record their heights. You find that the mean height of men is greater than the mean height of women in your sample. Your alternative hypothesis is "The mean height of men is different to the mean height of women." The null hypothesis is "There is no difference between the mean heights of men and women." You perform a hypothesis test and get a p-value of 0.3%. This means that assuming men and women have the same mean height, there would be a 0.3% chance of taking your sample and finding a result as extreme as the one you found.

A probability of 0.3% is low. It seems more sensible to **reject the null hypothesis** and conclude that there really is a difference in the mean heights of men and women in the population.

Before carrying out a hypothesis test, you must decide what percentage you wish to beat to reject the null hypothesis. This is called the **significance level** (α). This is the probability (**p-value**) that your test must beat for you to consider the result as **statistically significant**. It is common to use 5% as the significance level. This means that if the p-value is above 5%, you fail to reject the null hypothesis. If the p-value is below 5%, you reject the null hypothesis.

In Higher Applications of Mathematics there are three types of hypothesis test you will carry out as detailed in the table below. Chapter 9 will show you how to perform these tests. In this chapter you will focus on **interpreting** the output of the tests.

Name of the test	What it tests	Null hypothesis	Alternative hypothesis	Example
t-test	Whether the means of two populations are different.	"There is no difference between the means of population A and population B."	"There is a difference between the means of population A and population B."	Is there a difference in weights between Eastern and Western honeybees?
z-test for two proportions.	Whether the proportion of a categorical characteristic differs between two populations.	"There is no difference between the proportions of population A and population B."	"There is a difference between the proportions of population A and population B."	Is there a difference in the proportion of Eastern and Western honeybees with mites?
Correlation test	Whether a correlation exists between two variables within a population.	"There is no correlation between A and B in the population."	"There is a correlation between A and B in the population."	Is there a correlation between insecticide use and the number of honeybees?

Example 8.1

Michael is investigating whether women are more likely to have a library card than men. He takes a random sample of 16 women and finds 12 of them have a library card. He samples 17 men and finds 11 of them have a library card. That's 75% of the women, and around 65% of the men.

He wants to know whether these differences are statistically significant. He sets a significance level of $\alpha = 5\%$ and runs a hypothesis test. His results are:

```
      2-sample test for equality of proportions with continuity correction

data: c(12, 11) out of c(16, 17)
X-squared = 0.069761, df = 1, p-value = 0.3958
alternative hypothesis: greater
95 percent confidence interval:
 -0.2185868  1.0000000
sample estimates:
    prop 1      prop 2
 0.7500000   0.6470588
```

> When you complete a hypothesis test, look for the *p*-value.

> You will find out more about confidence intervals later in the chapter.

a   State Michael's null hypothesis.

b   State Michael's alternative hypothesis

c   Should Michael reject, or fail to reject, the null hypothesis? What would this mean for the research undertaken?

a   Michael's null hypothesis would be "There is no difference in the proportion of men and women who have library cards."

b   The alternative hypothesis is "A higher proportion of women have a library card than men."

c   Michael's significance level is 5%. His hypothesis test has returned a **p-value** of 39.58%. As 39.58% > 5%, Michael should **fail to reject** the null hypothesis. The conclusion Michael would make is that he does not have enough evidence to conclude that the proportion of women with library cards is higher than the proportion of men.

> The phrasing is "fail to reject the null hypothesis". You never say "accept the null hypothesis." Not being able to show that something is true does not mean you must accept that it is false.

A **p-value** tells you the probability of getting data as extreme as the data you have sampled **given the null hypothesis is true.** In a world where men and women are both equally likely to have library cards, what is the probability that if Michael repeated his sampling he would find data as extreme as the one he found above? In this case, the answer is a 39.58% chance.

Type I and type II errors

A **type I error** is when you reject the **null hypothesis** but in fact the null hypothesis is true.

A **type II error** is when you fail to reject the **null hypothesis**, but in fact the null hypothesis is false.

You are expected to be able to discuss the results of a hypothesis test in the context of the research question, and to discuss the possibility of type I and type II errors occurring and the implications of these.

A researcher is interested in knowing whether there is a difference in the average weights of honeybees in countries where certain pesticides are banned, compared with countries where the pesticides are routinely used.

A type I error would be the researcher concluding that there is a difference in the weights of honeybees between the two regions, when in fact the average weights in the populations are the same.

A type II error would be the researcher failing to reject the null hypothesis and being unable to show that there is a difference in the weights, when in fact there really was a difference between the mean weights in the population.

Example 8.2

A school needs to decide on whether to offer Higher Applications as an option next year. They conduct a survey of S4 pupils to see what proportion of them are interested in Higher Applications. They will offer the course if 25% of students are interested. The school's hypotheses are as follows:

$$H_0: p \leq 25$$
$$H_A: p > 25$$

> H_0 is used for the null hypothesis, and H_A is used for the alternative hypothesis.

Which of the following would be a type I error?
Which would be a type II error?

a   They offer the course when they shouldn't have.

b   They don't offer the course and they shouldn't have.

c   They offer the course when they should have.

d   They don't offer the course when they should have.

a   Offering the course means they have rejected the null hypothesis when they should not have. This is a **type I** error.

b   The second statement is not an error, it's good that they didn't offer the course if not enough students were interested.

c   The third statement is not an error either, the school offering the course when the students are interested is good.

d   This statement means they failed to reject the null hypothesis, but they should have. This is a **type II** error.

Exercise 8A

1   A factory owner owns a machine that produces an average of 35 components in an hour. She purchases a second machine, and she suspects that the new machine produces components at a faster rate. She samples the output of the new machine and finds that the average of the sample is 37 components an hour.

a State the null hypothesis.

b State the alternative hypothesis.

★ **2** A charity has a website with a donation page. On average, 0.5% of visits to the website result in a donation being made to the charity. A web manager at the charity changes the website to make the donate button 3 times larger. He wants to know whether this has changed the average amount of donations being made per web page visit. He looks at a sample of the data and finds the sample mean is 0.8% of visits results in a donation.

a State the null hypothesis.

b State the alternative hypothesis.

★ **3** Mary carries out a series of hypothesis tests. For her work she has set a significance level of $\alpha = 5\%$. For each result below, determine whether the null hypothesis should be rejected.

a $t = 3.886$, df = 17, p-value = 0.4

b $t = 0.0425$, df = 17, p-value = 0.4

c $t = 3.886$, df = 17, p-value = 0.03145

d $t = 3.886$, df = 17, p-value = 0.009

> ⚠ Statistical outputs will often contain more information than you need. When interpreting the results of a hypothesis test, first look for the p-value.

4 Adele believes her data may be correlated. She calculates the correlation coefficient to be 0.79. She runs a hypothesis test with a significance level of 5% and the following is the output:

```
Pearson's product-moment correlation

data: A and B
t = 6.6728, df = 27, p-value = 3.67e-07
alternative hypothesis: true correlation is not equal to 0
95 percent confidence interval:
 0.5943646 0.8963168
sample estimates:
      cor
0.7889968
```

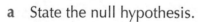

> ⚠ 3.67e-07 means "3.67 multiplied by ten to the power of negative 7."

a State the null hypothesis.

b State the alternative hypothesis.

c Should she reject or fail to reject the null hypothesis?

d Adele decides to reject the null hypothesis. In fact, there is no correlation between the variables in the population. What type of error is this?

★ **5** A company produces kettles that take on average 210 seconds to boil. It wishes to test whether a new kettle boils faster on average. The company's hypotheses are:

$H_0: \mu \geq 210$

$H_A: \mu < 210$

> The Greek letter mu (μ) represents the population mean.

Which of the following would be a type I error? Which would be a type II error?

a $\mu = 208$ and the company concludes that the new kettle has a lower mean boiling time.

b $\mu = 208$ and the company does not conclude that the new kettle has a lower mean boiling time.

c $\mu = 215$ and the company concludes that the new kettle has a lower mean boiling time.

d $\mu = 215$ and the company does not conclude that the new kettle has a lower mean boiling time.

6 A researcher investigating memory is testing the hypothesis that participants score a greater score on average in a memory test when they are given pictures in addition to the words. The researches conducts a test and gets the following result:

```
data: control_group and picture_group
t = -126.87, df = 996.73, p-value < 2.2e-16
alternative hypothesis: true difference in means is less than 0
95 percent confidence interval:
-Inf -7.926514
sample estimates:
mean of x    mean of y
 10.01610     18.04683
```

For the test, a significance level of $\alpha = 5\%$ was set.

a What is the null hypothesis in this situation?

b Should the null hypothesis be rejected?

c What conclusion could the researcher reach?

★ 7 Lisa has conducted a survey about the favourite flavour of ice-cream of students at her school. She wishes to see if there was a difference between boys and girls in their preference for vanilla. She conducts a test and gets the following result:

```
2-sample test for equality of proportions with continuity correction

data: c(47, 48) out of c(91, 109)
X-squared = 0.86723, df = 1, p-value = 0.3517
alternative hypothesis: two.sided
95 percent confidence interval:
-0.07262783 0.22486092
sample estimates:
    prop 1       prop 2
0.5164835    0.4403670
```

For the test a significance level of $\alpha = 5\%$ was set.

a What is the null hypothesis in this situation?

b Should the null hypothesis be rejected?

c What conclusion could Lisa reach?

d Suppose in fact there really is a difference in proportion between boys and girls and their favourite flavours of ice-cream in the population. What type of error would Lisa be likely to make?

★ 8 An insurer is considering whether to purchase an IT system from company A or company B. Using a sample of data, it conducts a hypothesis test to see whether there is a significant difference between the costs.

```
data: company_A and company_B
t = -328.97, df = 37.688, p-value < 2.2e-16
alternative hypothesis: true difference in means is not equal to 0
95 percent confidence interval:
-100.44023  -99.21129
sample estimates:
mean of x    mean of y
 147581.9     147681.8
```

For the test a significance level of $\alpha = 5\%$ was set.

a   What is the null hypothesis in this situation?

b   Should the null hypothesis be rejected?

c   What conclusion could the insurer reach?

> ⚠️ Think carefully about your answer to part **c**. Look at the values under "mean of x" and "mean of y."

Confidence intervals

You can take a random **sample** to make an estimate about a **population**. You can calculate the **means**, **proportions,** or **correlation coefficient** of your sample data, but your aim is to understand these measures in the **population**.

A **confidence interval** gives you a range of uncertainty for the parameter you are testing. The wider the confidence interval, the less certain you are of the population parameter. A 95% confidence interval means that if you repeated your study 100 times, 95 of those times the true population parameter would be contained in the confidence interval.

Imagine that 3% of honeybees have mites. A beekeeper samples some bees and finds that 2.8% of the bees have mites. She produces the following confidence interval: 1.9% to 4.1%. Her statistical output would look like this:

```
95 percent confidence interval:
 0.01903141 0.04075467
sample estimates:
     p
0.028
```

The beekeeper does not know that the true population proportion is 3%, but she can be 95% confident that the population proportion is between 1.9% and 4.1%.

Another beekeeper takes a different sample and finds that 3.1% of the bees have mites. His confidence interval is between 2.2% and 4.4%. He is 95% confident that the true population proportion is between those two percentages.

If 100 beekeepers took a similar sample, and produced their own confidence intervals, you would expect 95 of them to produce a confidence interval that contained 3%.

It's important to understand that a confidence interval is not a maximum or a minimum the value could be. When you produce a confidence interval you are likely to be one of the 95% who manage to capture the correct parameter (in this example, the proportion of mites). But there is a chance you are one of the unlucky 5%.

Why 95%?

The reason you choose a 95% confidence interval is because approximately 95% of values, in a normal distribution, will be within 2 standard deviations of the mean.

Example 8.3

Paul wants to know the average time it takes for his favourite takeaway to deliver food. He takes a sample of 100 deliveries and constructs a 95% confidence interval for the mean speed. This is his result:

```
95 percent confidence interval:
 24.456721  32 14574
```

a   Which of the following statements are true:

   i   If Paul took another sample of 100 deliveries then there would be a 95% probability that the sample mean would lie between 24.456721 and 32.14574.

   ii   95 of the deliveries in Paul's sample had a delivery time between 24.456721 minutes and 32.14574 minutes.

   iii   If Paul repeated his investigation over and over, taking more and more samples, then approximately 95% of the confidence intervals he produced would include the true average time it takes his favourite takeaway to deliver food.

   iv   There is a 95% probability that the true average time takes Paul's favourite takeaway to deliver food is between 24.456721 minutes and 32.14574 minutes.

   v   Paul can be 95% confident that the takeaway delivers food on average between 24.456721 minutes and 32.14574 minutes.

b   Paul takes a sample of a different takeaway, and produces the following confidence interval:

```
95 percent confidence interval:
 33.41321  35.35117
```

Paul says: "I am more certain about the true average delivery time for the second takeaway." Comment on what Paul says.

a   i   This is false. Confidence intervals tell you about the population parameter (in this case, the true average time it takes for the takeaway to deliver food), they do not make predictions about the results of other samples.

   ii   False. Confidence intervals tell you about the population parameter, they do not tell you about the distribution of your sample data.

   iii   This statement is true. You are 95% confident that your interval captured the population mean, so if you repeatedly sample the population you would expect 95% of the confidence intervals you produce to contain the true population mean.

   iv   This is false. The population mean (in this case, the true average time it takes for the takeaway to deliver food) either is between 24.456721 minutes and 32.14574 minutes or it isn't. Remember, the population mean is not a variable, it is a fixed fact about the population. The mean is what it is, and it does not make sense to say there's a 95% chance that it lies between two amounts. Either it does or it doesn't.

> **v** This is true. A 95% confidence interval suggests you can be 95% confident that the true population mean is in the interval. In this case, that the true average time to deliver the takeaway is between 33.41 and 35.35 minutes.
>
> **b** The second confidence interval is narrower, and therefore there is less uncertainty about the second takeaway's delivery times. Paul is correct.

Exercise 8B

1 Caitlyn collects a sample of data on the scores of football matches. Caitlyn conducts a hypothesis test on two variables: the number of goals scored by the boys' football team and the number of goals scored by the girls' team. She wants to know whether there is a significant difference in the average goals scored per match. For her work she has set a significance level of $\alpha = 5\%$. Her null hypothesis and alternative hypothesises are as follows:

$H_0 = \mu_1 - \mu_2 = 0$

$H_A = \mu_1 - \mu_2 \neq 0$

Where μ_1 is the **population** mean number of goals scored by the boys' football team, and μ_2 is the population mean number of goals scored by the girls' football team. Below is the result of her test:

```
t = 2.2647, df = 79.881, p-value = 0.02624
alternative hypothesis: true difference in means is not equal to 0
95 percent confidence interval:
0.0386956 0.5996023
sample estimates:
mean of boys   mean of girls
    1.702128       1.382979
```

a Explain why the null hypothesis is $H_0 = \mu_1 - \mu_2 = 0$.

b Should Caitlyn reject the null hypothesis?

c Is there a significant difference between the two football teams?

d Interpret the confidence interval of "0.038 695 6   0.599 602 3."

e Caitlyn says: "If I did another sample, there would be a 95% probability that the difference in the means in that sample would be between 0.039 and 0.6." Is this correct?

★ 2 Abdullah wants to test to see whether there is a correlation between people's scores on a reading test and on a memory test. Each participant is given a reading comprehension test and then tested on their ability to remember information from a video. Each activity is scored out of 100. Abdullah plots his data and creates a confidence interval. His significance level is 10%.

```
        Pearson's product-moment correlation

data: E and F
t = -1.1339, df = 27, p-value = 0.2668
alternative hypothesis: true correlation is not equal to 0
95 percent confidence interval:
 -0.5376875 0.1663051
sample estimates:
      cor
-0.2131952
```

    a State Abdullah's null and alternative hypotheses.

    b The Pearson's correlation value is -0.21. Interpret this value.

    c Should Abdullah reject or fail to reject the null hypothesis?

    d Interpret the confidence interval.

    e Abdullah says: "it is plausible that the true correlation in the population is zero." Is this correct?

3 Below are the outputs from two hypothesis tests:

Test 1

```
2-sample test for equality of proportions with continuity correction

data:  c(50, 60) out of c(100, 100)
X-squared = 1.6364, df = 1, p-value = 0.2008
alternative hypothesis: two.sided
95 percent confidence interval:
 -0.24719748  0.04719748
sample estimates:
prop 1  prop 2
   0.5     0.6
```

Test 2

```
2-sample test for equality of proportions with continuity correction

data:  c(500, 600) out of c(1000, 1000)
X-squared = 19.8, df = 1, p-value = 8.598e-06
alternative hypothesis: two.sided
95 percent confidence interval:
 -0.14438565  -0.05561435
sample estimates:
prop 1  prop 2
   0.5     0.6
```

    a For which test(s) should the null hypothesis be rejected?

    b For which test(s) is zero a plausible difference between the proportions?

    c For which test do you have a greater certainty about the population parameter?

★ **4** A local government is interested to know whether there is a difference in the number of smokers in one town compared to another. In an initial study, the council surveyed a random 250 residents in each town to identify what proportion had ever smoked a cigarette. The significance level was set to $\alpha = 10\%$. The results were as follows:

```
2-sample test for equality of proportions with continuity correction

data:  smokers out of residents
X-squared = 0.20793, df = 1, p-value = 0.6484
alternative hypothesis: two.sided
95 percent confidence interval:
 -0.0.9274977  0.05274977
sample estimates:
prop 1  prop 2
  0.18    0.20
```

The local government decided to take a wider sample. The results of their second sample and hypothesis test were:

2-sample test for equality of proportions with continuity correction

```
data: smokers out of residents
X-squared = 2.6992, df = 1, p-value = 0.1004
alternative hypothesis: two.sided
95 percent confidence interval:
 -0.049508816  0.004406695
sample estimates:
    prop 1       prop 2
 0.1789264    0.2014775
```

a   Did the second test give the local government more certainty about the true difference in proportions between the two towns?

b   Did the tests give different results about whether to reject the null hypothesis?

c   If there is no difference in the proportion of smokers between the two towns, but the local government concluded that there was, what type of error would this be?

d   A local government worker says: "The p-value dropped from one test to the next. If you repeated the process again, and this time took a much larger sample, you would find that there is a difference between the two towns." Comment on this.

5   Sophie is looking at the average cost of a piece of jewellery in two shops. Based on Sophie's sampling, in shop A the average sale is around £51, in shop B the average sale is around £124. She runs a hypothesis test and gets the following output:

```
           Welch Two sample t-test

data: C and D
t = -4.0736, df = 33.385, p-value = 0.0002686
alternative hypothesis: true difference in means is not equal to 0
95 percent confidence interval:
 -109.75300  -36.66079
sample estimates:
mean of x    mean of y
50.68966    123.89655
```

Which of the following statements are reasonable?

a   Sophie can be very confident that there is a difference in the average sale in the two shops.

b   Sophie cannot be very certain about what the true difference in average sale between the two shops is.

c   It is plausible that one shop has an average sale price of £100 less than the other shop.

d   It is plausible that one shop as an average sale price of £120 less than the other shop.

e   There is a 95% probability that the true difference between average sale prices is between around £37 and £110.

f   Sophie can be 95% confident that the true difference between average sale prices is between around £37 and £110.

Confounding variables

Confounding variables is a type of statistical error in which the effect of one variable is wrongly attributed to another. You will need to be able to discuss confounding variables when interpreting the results of a hypothesis test.

Example 8.4

Martha is reviewing data about the number of hospitalisations in the UK for severe sunburn. She also has data about the number of ice-cream sales that have been made.

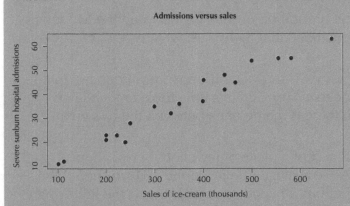

She finds that there is a strong linear relationship between the number of sunburn hospital admissions and the number of ice-creams sold. Martha says, "Ice-cream sales result in people getting sunburnt." Comment on what Martha says.

It is likely that there is another variable responsible for this relationship. In this case the number of days in summer with hot, sunny weather is likely have a strong, linear relationship with the number of sunburn hospitalisations and the number of ice-creams sold. Martha is confounding ice-cream sales with the number of hot, sunny days.

Exercise 8C

1   Harry is interested in finding out whether people who skateboard are more likely to describe themselves as being in good health, as opposed to people who don't regularly skateboard. Using a significance level of 5%, Harry runs a hypothesis test:

```
2-sample test for equality of proportions with continuity correction

data: c(220, 800) out of c(250, 1000)
X-squared = 8.0007, df = 1, p-value = 0.004676
alternative hypothesis: two.sided
95 percent confidence interval:
 0.03020026 0.12979974
sample estimates:
prop 1  prop 2
 0.88    0.80
```

a   State Harry's null and alternative hypothesis.

b   Can he reject or fail to reject the null hypothesis?

c Give an example of a variable that Harry risks confounding with skateboarding in his analysis.

2 A researcher is considering whether mothers are less likely to give birth to healthy babies during their third pregnancies. She analyses some data and finds that first and second babies are more likely to be born healthy compared to babies born third. Another researcher criticises the analysis, saying a variable has been confounded. Suggest what this variable could be.

3 Men are more likely to be smokers than women. Explain why this fact might be important if you wanted to investigate the difference in life expectancy between males and females.

★ 4 A supermarket wishes to know whether playing Christmas music leads to shoppers spending more money on average. The supermarket plays Christmas music between the final week of November and Christmas day.

a Explain why comparing a sample of customer shopping spends taken in December with a sample taken in July would be a poor way to test this hypothesis.

b The supermarket has two stores in Glasgow. The area manager decides to play Christmas music in one of the stores and play no music in the other store. She will then sample both stores and look for a significant difference in the average customer spend. After doing this the area manager concludes that playing Christmas music does increase the average amount a shopper spends in a shop. Give 3 reasons why this conclusion might not be accurate.

- I can interpret the statistical output of a hypothesis test. ★ Exercise 8A Q2
- I can explain what the null hypothesis and alternative hypothesis are. ★ Exercise 8A Q3, 8
- I can understand the difference between a type I and type II error. ★ Exercise 8A Q5, 7
- I can interpret confidence intervals. ★ Exercise 8B Q2
- I can compare two confidence intervals and discuss which one gives more certainly about a population parameter. ★ Exercise 8B Q4
- I can discuss confounding variables. ★ Exercise 8C Q4

9 Performing Hypothesis Tests

This chapter will show you how to:

- fit and interpret a linear model by applying code in R
- make predictions based on a linear model
- perform and interpret a correlation test
- perform and interpret *t*-tests and paired *t*-tests
- perform and interpret *z*-tests for proportions
- perform and interpret a hypothesis test for your project.

You should already know:

- how to interpret the correlation coefficient for bivariate data
- how to produce scatter plots in RStudio
- how to interpret the results of a hypothesis test
- how to interpret a confidence interval.

Performing hypothesis tests

Hypothesis tests are performed on a **sample**. The result of the test will help you make a conclusion about the **population**.

For example, if you take a sample of grey squirrels and red squirrels and measure their weight, you might find that grey squirrels are heavier on average **for that sample**. You need to perform a hypothesis test before you can conclude anything about whether the **population** of grey squirrels is heavier than the **population** of red squirrels on average.

The hypothesis tests in this chapter can only be performed when the sample is chosen randomly from the population and the data in the sample are independent. A correlation test can only be performed on continuous data that are approximately normally distributed. In t-tests the differences between both groups must be approximately **normally distributed**.

Linear models

You can calculate the **correlation coefficient** of a set of **bivariate** data. Usually your data will be a sample and your aim is to understand whether there is a correlation in the **population**.

Bivariate means "two variables". For example, a dataset containing the salaries and heights of employees in a company would be bivariate – the two variables are salaries and heights.

The correlation coefficient, referred to as "*r*", is a measure of correlation from −1 to 1. An *r* value close to −1 would suggest a strong, negative linear relationship. An *r* value close to 0 suggests either a weak, or no linear relationship. An *r* value close to 1 suggests a strong linear relationship exists between the two variables.

Recall from Chapter 8 that hypothesis tests require a **null** and an **alternative hypothesis**. The hypotheses for a correlation test are summarised below.

Name of the test	What it tests	Null hypothesis	Alternative hypothesis	Example
Correlation test	Whether a correlation exists between two variables within a population.	"The is no correlation between A and B in the population."	"The is a correlation between A and B in the population."	Is there a correlation between insecticide use and the number of honeybees?

Hypothesis tests produce a *p*-value. In a world where the null hypothesis is true, what is the probability of taking a sample and finding a correlation this extreme? If the *p*-value is low (usually less than 5%), then you reject the null hypothesis and conclude that a correlation does exist between the two variables in the population.

R code for linear regression

The following R code can be used to calculate the linear regression.

Fit a linear model to data – `lm(Y~X)`.

Add the linear model to a scatter plot – `abline(lm(Y~X))`

> Note that the **y** variable comes first here.

Perform a correlation hypothesis test – `cor.test(X, Y)`

Make a prediction based on a linear model – `predict(lm(Y ~ X), newdata=data.frame(X=C),interval = "pred")`.

> This finds the predicted value of *y* when *x* is equal to *c* (some constant).

Example 9.1

The dataset "91.csv" contains the prelim results, and final assessment results, for a sample of mathematics students.

a Import the data into RStudio and produce a scatter plot.

b Form a subjective impression.

c Fit a linear model to the data, and interpret the gradient and intercept parameters.

d Perform a correlation hypothesis test and interpret the result.

e Interpret the confidence interval.

f Predict the final result for a student who scored 70 on the prelim.

(continued)

a The following code will import the data and produce the plot:

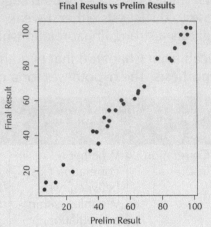

Final Results vs Prelim Results

```
1  #Import Data
2  assessment_data <- read.csv("91.csv")
3  attach(assessment_data)
4  names(assessment_data)
5
6  #Produce the Plot
7  plot(Prelim, Final,
8      main="Final Results vs Prelim Results",
9      xlab="Prelim Result",
10     ylab="Final Result",
11     pch=19)
12
13  |
```

b The question wants you to comment on whether there appears to be a linear relationship, and if so what kind. There appears to be a positive, linear relationship in this data.

c The following code will fit a model to the data:

```
13
14  #Producing a linear model:
15
16  lm(Final ~ Prelim)
17  |
```

```
> lm(Final ~ Prelim)

Call:
lm(formula = Final ~ Prelim)

Coefficients:
(Intercept)    Prelim
      1.417     1.005
```

This output is telling you that the equation for the linear model is:

Final = 1.417 + 1.005 × Prelim

The intercept value (1.417) is the score you would expect a student who scored zero on the prelim to achieve in the final. The gradient (1.005) tells you that as the prelim mark increases by 1 mark, the final assessment increases by 1.005 marks.

To add the linear model to the scatter plot, use this code:

Final Results vs Prelim Results

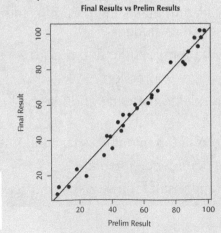

```
17
18  #Add to the plot
19  abline(lm(Final ~ Prelim))
20  |
```

d Before performing a hypothesis test, you state your two hypotheses:

Null hypothesis	Alternative hypothesis
"There is no correlation between A and B in the population."	"There is a correlation between A and B in the population."

A correlation hypothesis test is performed using the code:

```
#Hypothesis Test

cor.test(Prelim, Final)
```

```
> cor.test(Prelim, Final)

        Pearson's product-moment correlation

data: Prelim and Final
t = 39.77, df = 27, p-value = < 2.2e-16
alternative hypothesis: true correlation is not equal to 0
95 percent confidence interval:
 0.9819088 0.9960842
sample estimates:
      cor
0.9915722
```

> This is the *p*-value in scientific notation. Remember having a negative index, like −16, means the value is small.

> The test will generate a confidence interval. You can be 95% confident that the true population correlation is between 0.981 9088 and 0.996 0842.

> The correlation coefficient, *r*.

The value under "cor" is the correlation coefficient, *r*. In this case 0.991 5722 suggests a strong, positive, linear relationship.

Remember that *r* is measured between −1 and 1, with 1 being the strongest, positive relationship.

The *p*-value (see Chapter 8) is the probability of obtaining a correlation coefficient as extreme as this if there was no correlation in the population. In this case the *p*-value is very close to zero, so you can reject the null hypothesis and conclude that the true correlation in the population is not zero.

e You can be 95% confident that the true population correlation is between 0.982 and 0.996.

f You can make a prediction based on your linear model with this code:

```
#Predicting
predict(lm(Final ~ Prelim), newdata=data.frame(Prelim=70), interval = "pred")
```

This is a reasonable prediction to make because 70 is within the range of the data. The output is:

```
> predict(lm(Final ~ Prelim), newdata=data.frame(Prelim=70), interval = "pred")
      fit      lwr      upr
1 71.79772 63.90516 79.69029
```

The estimated result would be 72 in the final, however the true value is likely to be between 64 and 80. R will usually round values to 5 decimal places. Often it makes sense to round to two or three significant figures.

Exercise 9A

For this exercise import the file "9A.csv" into RStudio. See Chapter 7 for a reminder on importing data.

★ 1   A researcher used a sample of women between the ages of 20 and 30. She recorded their age, height in cm, and hand span in cm. Each participant in the study was given a quiz on general knowledge (scored out of 20) and was challenged to throw a basketball into a basket as many times as possible in 30 seconds.

    a   Produce a scatter plot for age and quiz score.

    b   Calculate the correlation coefficient.

    c   Produce a scatter plot for height and baskets scored.

    d   i   Perform a hypothesis test for the correlation between height and baskets scored.

       ii   Should the null hypothesis be rejected?

    e   Interpret the confidence interval.

    f   Fit a linear model to the data for height and baskets scored.

    g   Display the line on your scatter plot, and state the equation for this line.

    h   Interpret the gradient and intercept parameters for your linear model fitted in part **f**.

    i   Predict the number of baskets scored for a women with a height of 184 cm.

    j   Interpret the prediction interval for part **i**.

    k   Explain why it would be inappropriate to predict the number of baskets scored for a woman of height 240 cm.

    l   Morgan says: "I have tested the correlation between hand span and baskets scored, and there's a correlation. People with bigger hands are better at basketball." Comment on this.

    ⚠️ What variables might Morgan be confounding here?

2   The dataset contains 4 pairs of data: A1 and A2, B1 and B2, C1 and C2, D1 and D2.[*]

    a   For each pairing calculate the:

       i   correlation

       ii   equation of the linear model.

    b   Comment on the similarity and differences of these four pairs of data.

    c   For each pair, produce a scatter plot.

    d   After plotting the data, comment on the suitability of a linear model.

---

* Anscombe, F. J. (1973). "Graphs in Statistical Analysis". American Statistician. 27 (1): 17–21. doi: 10.1080/00031305.1973.10478966

3 The dataset has data relating the mass (in kg) of men who took part in a study, along with the number of cigarettes they smoke in a day.

   a Produce a scatter plot of mass and cigarettes smoked.

   b Calculate the correlation coefficient. Interpret the correlation.

   c Perform a hypothesis test for the correlation. Should the null hypothesis be rejected?

   d Adam says: "When doing a correlation test, if you get a high correlation coefficient you always get a low p-value." Is this correct?

★ 4 The dataset has data showing the scores students achieve in their English and History tests.

   a Produce a scatter plot for English and History score.

   b Calculate the correlation coefficient.

   c Perform a hypothesis test for the correlation. Should the null hypothesis be rejected?

   d Interpret the confidence interval.

   e Find the equation of the linear regression line.

   f Interpret the gradient and intercept parameters for your linear model fitted in part **e**.

   g Predict the History score for a student scoring 8 on their English test. Making reference to the prediction interval, comment on how certain you are about your prediction.

   h Brooke says: "When doing a correlation test, if you get a low correlation coefficient you always get a high p-value." Is this correct?

5 Import the file named "9 Scotland Pollution.csv" into R. This dataset contains emissions readings from different water sources in Scotland, with a description of the type of water source and the year the sample was collected.

   a Produce a scatter plot for the Emissions and Year.

   b Form a subjective impression.

   c Calculate the correlation coefficient.

   d Perform a hypothesis test for the correlation. Should the null hypothesis be rejected?

   e Interpret the confidence interval. Is there a relationship between the year a sample was taken and the amount of pollution?

T-tests

With **numerical data** you often want to compare the average of two groups. For example, which drug is more effective at reducing heart-attacks, drug A or drug B?

The hypothesis for comparing the mean between two groups is called a **t-test**. The hypothesis for a t-test are summarised below:

Name of the test	What it tests	Null hypothesis	Alternative hypothesis	Example
t-test	Whether the means of two populations are different.	"There is no difference between the means of population A and population B."	"There is a difference between the means of population A and population B."	Is there a difference in weights between Eastern and Western honeybees?

There are two t-tests: paired and unpaired. A **paired** t-test is used when the item is tested in both groups. For example, a person's blood pressure is measured before and then after taking some medication. When making a comparison between the before and after measurements, as it is the same people in both groups, a paired t-test is required.

An **unpaired** t-test is used when the items in both groups are **independent**. For example, measuring the blood pressure of a group of people living in a large city and a separate group of people living in a rural area and then comparing the blood pressures of each group. As the people in each group are different, an unpaired t-test is required.

Below is the R code required to run a t-test:

> ## *T*-tests in RStudio
>
> Unpaired t-test – `t.test(X, Y)`
>
> Paired t-test – `t.test(X, Y, paired=TRUE)`
>
> X and Y are the variable names. You will replace the X and Y with the names of the columns in your spreadsheet.

> ## Example 9.2
>
> A market researcher asks a sample of people to rate how much they like a company out of 10. They then show them an advert for the company. Afterwards, the participants in the sample are asked to rate the company again. The data collected is in the file "92.csv".
>
> a   Import the data into RStudio. What type of data are the variables?
>
> b   Represent this data on a suitable diagram.
>
> c   Produce statistics showing the measure of spread and dispersion.
>
> d   Is there a statistically significant difference between the scores taken before watching the advert and afterwards?
>
> e   Interpret the confidence interval.

See the file "92.R" for all the R code required to complete this example. The file "92.csv" contains the data to be imported.

a This is discrete, numerical data.

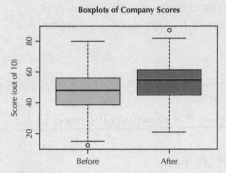

b
```
 6  #b
 7  boxplot(Score.Before, Score.After,
 8          main="Boxplots of Company Scores",
 9          ylab="Score (out of 10)",
10          names=c("Before", "After"),
11          col=c("cadetblue3", "hotpink") #Colours are optional of course
12  )
```

c First you check whether the data are normal or not, using the hist() function.

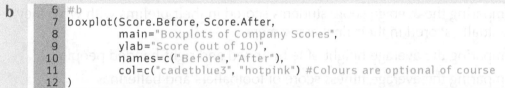

As both datasets are normal, the mean and standard deviation are the correct measures.

Data	Mean	Standard deviation
Before	47.8	13.9
After	53.5	14.9

d
```
> t.test(Score.Before, Score.After, paired = TRUE)

        Paired t-test

data:  Score.Before and Score.After
t = -11.497, df = 58, p-value = < 2.2e-16
alternative hypothesis: true difference in means is not equal to 0
95 percent confidence interval:
 -6.726251 -4.731376
sample estimates:
mean of the differences
            -5.728814
```

(continued)

As the *p*-value is less than 5%, you can reject the null hypothesis and conclude that there is a difference between the two means. **You used a paired *t*-test as the same people were present in each group.**

e   You can be 95% confident that the true difference between the means is between −6.73 and −4.73. In other words, one of the groups is between 4.73 and 6.73 lower on average than the other.

Exercise 9B

1   For each of the following examples, state whether the *t*-test would need to be paired or unpaired:

   a   Comparing the average salaries of men and women.

   b   Comparing the average score students scored in their prelims with what they eventually scored in their final assessment.

   c   Comparing the average height of left-handed and right-handed people.

   d   Comparing the average fitness score of footballers and ballerinas.

   e   Comparing the average fitness score of footballers before and then after a new fitness programme is introduced.

⊕ Questions 2, 3 and 4 require the dataset "9B.csv". Import this dataset into RStudio now.

★ 2   The dataset contains data for left-handed and right-handed people who took part in a test to check their reaction time. Participants were given a score out of 30, with 30 being the fastest reaction time.

   a   Display the data in a suitable diagram.

   b   Produce summary statistics for the two groups.

   c   Form a subjective impression – is there a difference?

   d   Perform a suitable hypothesis test and interpret the result.

   e   Interpret the confidence interval.

3   The dataset has information about systolic blood pressure (mmHg) taken before and then after watching a stressful video.

   a   Display the data in a suitable diagram.

   b   Produce summary statistics for the two groups.

   c   Form a subjective impression – is there a difference?

   d   Perform a suitable hypothesis test and interpret the result.

   e   Interpret the confidence interval.

4   A charity fundraising team reviews a sample of donations it received over the past year. Donations (in £s) received without Gift Aid are recorded under the column named "Donation". Donations where the giver was registered for Gift Aid are recorded under the column "Donation_Gift_Aid". The fundraising team want to know whether Gift Aid registered donations were different on average to non-Gift Aid donations.

   a   Display the data in a suitable diagram.

b   Produce summary statistics for the two groups.

c   Form a subjective impression – is there a difference?

d   Perform a suitable hypothesis test and interpret the result.

e   Interpret the confidence interval.

5   Import the file named "9 Scotland Pollution.csv" into R. This dataset contains emissions readings from different water sources in Scotland, with a description of the type of water source and the year the sample was collected.

a   Produce a boxplot for the columns named "River Emissions" and "Coastal Emissions."

b   Display the same data in histograms.

c   Comment on the distribution of the data for River and Coastal emissions.

d   Adam performs a *t*-test to compare the difference between River and Coastal emissions. What assumption is Adam making by performing this test?

Z-tests for two proportions

With **categorical data** you often want to compare the proportion of two groups. For example, which library, the library in town or the mobile library touring the villages, has a higher proportion of non-fiction books?

The hypothesis test for comparing the proportion between two groups is called a **z-test for two proportions**. The hypotheses for a *z*-test for two proportions are summarised in the table:

Name of the test	What it tests	Null hypothesis	Alternative hypothesis	Example
z-test for two proportions.	Whether the proportion of a categorical characteristic differs between two populations.	"There is no difference between the proportions of population A and population B."	"There is a difference between the proportions of population A and population B."	Is there a difference in the proportion of Eastern and Western honeybees with mites?

The R code is as follows:

Z-test for two proportions in RStudio

The R code for a *z*-test for two proportions is: `prop.test(x = c(a, b), n = c(n1, n2))`.

This code will compare the proportion $\frac{a}{n1}$ with the proportion $\frac{b}{n2}$.

For example, if you wanted to compare the proportion "8 out of 10" with the proportion "17 out of 20", you would type: `prop.test(x = c(8, 17), n = c(10, 20))`.

The first bracket contains both numerators, the second bracket contains both denominators.

Example 9.3

Import the data file "Ice_cream.csv" into RStudio. This file contains data about the gender of students sampled in a study, and their favourite ice-cream flavour.

a   Produce a contingency table for Gender and Favourite Ice-cream.

b   Calculate the proportion of females who preferred vanilla ice-cream, and calculate the proportion of males who preferred vanilla ice-cream.

c   Perform a z-test for two proportions and interpret the result.

d   Interpret the confidence interval.

Open the file "93.R".

a   After importing the data, you can produce a table using `table(gender, ice_cream)`.

```
> table(gender, ice_cream)
          ice_cream
Gender   Chocolate  Strawberry  Vanilla
  Female     32         29        48
  Male       15         29        47
```

b   There are $\dfrac{48}{109} \approx 44\%$ of females who prefer vanilla, and $\dfrac{47}{91} \approx 52\%$ of males who prefer vanilla.

c   You can use the following R code: `prop.test(x =c(48, 47), n=c(109,91))`.

```
        2-sample test for equality of proportions with continuity correction

data:  c(48, 47) out of c(109, 91)
X-squared = 0.86723, df = 1, p-value = 0.3517
alternative hypothesis: two.sided
95 percent confidence interval:
 -0.22486092 0.07262783
sample estimates:
   prop 1      prop 2
 0.4403670   0.5164835
```

As the p-value is above 5%, you would fail to reject the null hypothesis. You cannot conclude that there is a statistically significant difference between the proportion of male and female students who prefer vanilla ice-cream.

Notice that running this test calculates the proportions for you; they appear under "prop 1" and "prop 2" and match your answers. It is a good idea to calculate these proportions yourself, like you did in part **b**, to check that you ran the proportion test correctly.

d   You can be 95% confident that the true difference in proportion between males and females is between −0.22 and 0.07. Zero is a plausible number within this range, which supports your decision not reject the null hypothesis.

Exercise 9C

1 Scott is looking at data about patients involved in a clinical trial:

Treatment	Blood type O negative	
	Yes	No
Successful	30	77
Unsuccessful	60	21

a Calculate the proportion of patients whose treatment was successful who had O negative blood.

b Calculate the proportion of patients whose treatment was not successful who had O negative blood.

> When you are given a table of information, you do not need to import any data into RStudio. Just use the R code for performing the z-test for two proportions.

c Perform a z-test for two proportions and interpret the result.

d Interpret the confidence interval.

2 A researcher is reviewing hospital data concerning patients who were admitted after being involved in a car accident. The data are summarised in the table:

Seatbelt	Outcome	
	Survived	Died
Yes	159	22
No	56	85

a Calculate the proportion of people who wore seatbelts who survived the car accident.

b Calculate the proportion of people who were not wearing seatbelts who survived.

c Perform a z-test for two proportions and interpret the result.

d Interpret the confidence interval.

⬇★ 3 Import the data file named "9C.csv".

Harry has sampled books from his school library and noted the gender of the author (under the column "Gender") and whether the book written was fiction or non-fiction (under the column "Type").

a What type of data has Harry collected?

b Display the data using suitable diagrams.

c Produce a contingency table for Gender and Type.

d Calculate the proportion of fiction books with a female author.

e Calculate the proportion of non-fiction books with a female author.

f Perform a z-test for two proportions and interpret the result.

    g Interpret the confidence interval.

    h Harry says: "In Scotland, women are no more likely to write fiction books than non-fiction books". Give two reasons why Harry cannot make this claim on the basis of his current data.

4 The file "9C.csv" also contains data from a sample of the Scottish population. Participants were asked their age and asked whether they are in good health. The data are represented in the spreadsheet files as follows:

Number used	Age	Health
1	0–20	Very good
2	21–45	Good
3	46–64	Poor
4	65+	Very poor

    a Produce pie charts displaying the **percentage** of participants by age and by health.

    b Produce a contingency table for age and health.

    c Perform a z-test for two proportions to compare the proportion of 21- to 45-year-olds who describe themselves as being in good or very good health with the proportion of 65+ year-olds who describe their health as good or very good.

    d Interpret the result of the hypothesis test and the confidence interval.

5 Import the file named "9 Scotland Pollution" into R. This dataset contains emissions readings from 2929 different water sources in Scotland, with a description of the type of water source and the year the sample was collected.

    a Produce a table showing the number of different sources that were sampled, using the "Type" column.

    b Produce a table showing the percentage of each different source sampled.

    c Display this information in a suitable diagram.

    d Adam decides to run a proportion test to compare the proportion of Lochs sampled with Rivers. Here is his output:

```
> prop.test(x = c(960, 862), n = (2929, 2929))

        2-sample test for equality of proportions with continuity correction

data:  c(960, 862) out of c(2929, 2929)
X-squared = 7.4954, df = 1, p-value = 0.006186
alternative hypothesis: two.sided
95 percent confidence interval:
 0.009424107 000.057492930
sample estimates:
    prop 1      prop 2
 0.3277569   0.2942984
```

    Adam says: "There is a significant difference in proportion between lochs and rivers in Scotland." Is this correct?

Project

As part of your project you will need to run a hypothesis test. You have sourced some data that is a **sample** of a **population**. You have formed a subjective impression about that data. Now you will use a hypothesis test to come to a conclusion about your population.

You will either be exploring whether a correlation exists between two variables, whether there is a difference between the means of two groups, or whether there is a difference between the proportions of a characteristic between two groups. Think carefully about your data type and what you compared in your subjective impression.

Example 9.4

Emilija is completing her project. She is interested in finding out whether red Smarties® are less likely to be found in a packet of Smarties than green ones.

Emilija has collected a sample of 500 Smarties and recorded their colours. As this is categorical data, she will use a z-test for two proportions.

```
        2-sample test for equality of proportions with continuity correction

data: c(105, 96) out of c(500, 500)
X-squared = 0.39851, df = 1, p-value = 0.5279
alternative hypothesis: two.sided
95 percent confidence interval:
 -0.0336638 0.0696638
sample estimates:
prop 1   prop 2
 0.210    0.192
```

The confidence interval shows that Emilija can be 95% confident that the true difference in proportions between red and green Smarties is between -0.03 and 0.07. Zero is within this range which tells Emilija that this is a plausible value; the difference could plausibly be zero.

The p-value is greater than 5%. Emilija's null hypothesis is that there is no difference in proportions between green and red Smarties, and she cannot reject this as 52% > 5%. This supports what you saw in the confidence interval; you cannot say that the difference is not zero.

Emilija notes that this matches with her subjective impression. There was no clear modal colour of Smartie as the proportions of Smarties were very similar, and this was seen clearly in the bar charts she produced.

After you have found your dataset, introduced it, and formed a subjective impression, you need to analyse and interpret your data using a hypothesis test.

Answer these questions:

1   What type of data are you investigating?

2   What is the correct hypothesis test to use for this data type?

3 Perform either a *t*-test, paired *t*-test, *z*-test for proportions or find a regression line, depending on your answer to Question 2.

4 State the null and alternative hypothesis for your test.

5 What is the confidence interval for your data? If you are exploring a correlation then find the correlation coefficient.

6 Interpret the confidence interval. Is zero a plausible value? If you are exploring correlation, comment on the strength of the linear relationship.

7 What is the *p*-value for your hypothesis test?

8 What is your null hypothesis, and should you reject or fail to reject it?

9 Comment on how your hypothesis test relates to your measure of location and dispersion, or for categorical data the proportions and the sample size. Did it match with your subjective impression?

10 Comment on how your hypothesis test relates to the graphical displays you have produced. Did it match with your subjective impression?

11 What assumptions have you made when conducting your project? Can you validate any of them?

- I can fit and interpret a linear model. ★ Exercise 9A Q1, 4
- I can make predictions based on a linear model. ★ Exercise 9A Q1, 4
- I can correctly use R code to perform hypothesis tests. ★ Exercise 9A Q4, 9B Q2, 9C Q3
- I can perform and interpret a correlation test. ★ Exercise 9A Q1, 4
- I can perform and interpret *t*-tests and paired *t*-tests. ★ Exercise 9B Q2
- I can perform and interpret *z*-tests for proportions. ★ Exercise 9C Q3
- **I can perform and interpret a hypothesis test for my project.**

10 Writing Your Project

This chapter will show you how to:

- structure your project
- match your data type to the correct graphical display and hypothesis test
- conclude your project
- present your project.

You should already know:

- the content covered in Chapters 4–9.

Project structure

Your project will be a report that includes the results from the statistical software that you are using, including tables, charts, graphs and the result of statistical calculations. The report should have the following sections:

- **Introduction**
- **Subjective impression**
- **Analysis and interpretation**
- **Conclusion.**

You will also be marked on the **presentation** of your project.

These sections match up with the chapters of this book as follows:

Section	Chapter
Introduction	6
Subjective impression	7
Analysis and interpretation	9
Conclusion and presentation	10

Chapters 6, 7 and 9 each have a project section, with questions to help you think about your project. Completing these sections will help you gather all the information you need to write up your project in a 2000-word assignment.

When completing your project, you should always think about your research question. Your **introduction** is where you state clearly what your research question is and provide background information about the area it relates to.

The **subjective impression** is about generating graphical displays and statistics to help you form an initial view about what the answer to your research question might be.

You will complete your hypothesis test in the **analysis and interpretation** section of your project. It is important that you are clear about what your research question is and that your hypothesis test is attempting to answer it.

Finally, in your **conclusion** you need to state clearly the answer to your research question and make a connection between your conclusion and your graphical displays, descriptive statistics and hypothesis test.

The importance of data type

You should say what type of data (**categorical**, **numerical**, **bivariate**) you are investigating in your introduction. Your data type will inform what kind of graphical displays you produce and what type of tests you perform.

Exercise 10A

1 Rodney is investigating categorical data. What type of graphical display would you expect to see in his project?

2 Lizzie ran a correlation hypothesis test as part of her project. What data type is she investigating?

3 Graeme is investigating whether men have a higher blood pressure than women.
   a What type of data is Graeme investigating?
   b What hypothesis test would you expect to see in his project?
   c What type of graphical display would be appropriate?

4 Alison's project contains a proportion test. What is her data type and what graphical display might you expect to see in her project?

5 Cassie is investigating whether there is a link between a person's height and a person's earnings. She produces a scatter plot for her project. She decides to run a **paired** *t*-test as her hypothesis test, saying: "It's important that I use a paired test as the groups are not independent – it's the same people in each group!". Explain Cassie's mistake.

6 Alex's project contains boxplots.
   a What data type is she investigating?
   b What hypothesis test(s) could be used?
   c What descriptive statistics could be used?

7 Matthew has a pie chart in his project.
   a What data type would this relate to?
   b What descriptive statistics could be used?
   c What hypothesis test(s) could be used?

Presentation

Your report should be easy to read. You will be marked on the presentation of your project as well as its content. A friend should be able to read your project and understand what your question is, what you did and what your conclusion was.

Advice	Example(s)
Introduce your graphical displays with text.	"Below is a table showing the different flavours of ice-cream and the percentage of participants who favoured them. This table is represented in the pie-chart labelled Figure 3." <table><tr><th>Chocolate</th><th>Mint choc chip</th><th>Strawberry</th><th>Vanilla</th></tr><tr><td>30.952 38</td><td>30.952 38</td><td>9.523 81</td><td>28.571 43</td></tr></table> Figure 3:
Introduce your descriptive statistics with text.	"Because I concluded that the data of both groups are normal, I found the mean and standard deviation for both the men's weights and the women's weights. These statistics can be seen below in Table 2."
Give your sections appropriate headers	"Introduction" "Sourcing my data" "Hypothesis tests"
Ensure the report flows and is linked to the context	"Overall my subjective impression is that there is a correlation between hand span and height. I will now investigate whether the correlation is statistically significant by performing a correlation test."
Include your working, saying when you used a spreadsheet or a calculator.	"I used a spreadsheet to count the number of times 'Vanilla' was listed as a favourite flavour."

Making your project flow can be achieved with a technique called "signposting." Signposting is when a writer guides a reader through an essay or project by either briefly summarising what has already been covered, or explaining where the project will go next. Useful words and phrases include:

- "I will now …"
- "However, …"
- "Having first looked at …, I will now …"

- "Initially ..."
- "Next ..."
- "Finally ..."
- "We have already seen ... So now ..."

Ask a friend to read your project and ask them to tell you if they ever feel lost or if it's not clear what you are doing. Any sections they highlight will benefit from signposting.

You must also include a bibliography and an appendix of data. Your bibliography is where you cite any other sources you have used in producing your project. For example, the website where you downloaded your dataset, or a website that you used when providing background information in your introduction. If you took data from a book, include the name of the book, the author, the year it was published and the page number used.

Bibliography example
Bibliography

Data sourced from Sheffield University Datasets: https://www.sheffield.ac.uk/mash/statistics/datasets.

Additional data from *Statistics* by Hays, W., 2008. p.55.

It is important that you include a table showing the data you have used in your project. Put this at the end of your report as an **appendix.** If your dataset is very large, just include a table showing the first 10 rows. It is important to include an appendix as this will help the examiner understand your dataset and interpret your report.

Appendix example
Appendix

The table below contains the first 10 rows of my dataset:

Year	Type	Emissions	Units	River emissions	Coastal emissions
1998	Rivers	1251.2	$kg\,a^{-1}$	1251.2	118.3
1998	Rivers	1228.8	$kg\,a^{-1}$	1228.8	2917.3
1998	Rivers	826.7	$kg\,a^{-1}$	826.7	1945.1
1998	Rivers	739.2	$kg\,a^{-1}$	739.2	1919.6
1998	Rivers	488.1	$kg\,a^{-1}$	488.1	1348.4
1998	Rivers	333.9	$kg\,a^{-1}$	333.9	1306.7
1998	Rivers	324.4	$kg\,a^{-1}$	324.4	1291.2
1998	Rivers	306.7	$kg\,a^{-1}$	306.7	1047.9
1998	Rivers	301.2	$kg\,a^{-1}$	301.2	882.7
1998	Rivers	230.3	$kg\,a^{-1}$	230.3	513.7

The word count for the project is 2000 words. You will be penalised if your project is over 2200 words. There is no minimum word count.

Remember that your project will be viewed in **black and white**. Make sure any graphical displays can be interpreted without colour.

Sources of data

You can source your own data by doing your own **sampling**. For example, you could do a sample of S1 students in your school. Creating your own data can be time consuming, but the advantages are that you are in control of how the data are entered into a spreadsheet and so you can make sure the format is correct.

You can source data from the internet. You may need to adjust any spreadsheets to make them usable. See Chapter 4 for practice on this. Below is a list of websites to source data from:

Name	URL
CORGIS	https://corgis-edu.github.io/corgis/csv/
R Packaged Datasets	http://www-eio.upc.edu/~pau/cms/rdata/datasets.html
Our World In Data	https://ourworldindata.org/
Earth Data	https://earthdata.nasa.gov/

Using a search engine is another useful way to find data. Here are some tips for finding good datasets online:

- Include "csv" within your search term. This means you are more likely to find datasets that can be easily imported into RStudio.

- If you have trouble finding an easy to use dataset, try searching "beginner friendly datasets." There are plenty of datasets that have been put together for people who are new to statistical analysis. These datasets are less likely to need any tidying up by you.

- Look for websites that have no reason to give you biased data. Commercial websites that have something to sell, or a charity's website, may have data tailored to suit their aims. Websites that are independent can be more reliable sources of data.

Searching online example
Folk music

If you are interested in folk music, then consider searching "Folk music dataset beginner".

If you do this, you are likely to learn about Eric Foxley's Music Database. It contains scores of over 1000 folk melodies. You could take a random sample of the scores to investigate the time signatures of folk melodies, or compare the proportions of folk melodies in the key of G with the key of D.

Football

If you are interested in football, consider searching "football csv." There are many football datasets to be found online, and they are regularly updated. Possible investigations include "Are teams more likely to score more goals at home or when away?", or "Which European league produces the most goals per season on average?".

Conclusion

In your conclusion you need to summarise what you have done and give a clear answer to your research question. In doing so you must connect your conclusion to the statistics you have produced and the graphical displays you have produced. **You must refer to your statistics, graphical displays and your hypothesis test in your conclusion.** You will have already discussed these things in your project, but you must refer to them **again** in your conclusion.

⚠ Your conclusion can be negative. There is nothing wrong with finding out that your hypothesis was incorrect.

Example 10.1

Mike is summarising his project. Mike gave a sample of students a memory test in which they had 1 minute to memorise a list of 20 words related to cars. He then tested the students to see how many words they could remember. His hypothesis was that male students would score higher than female students.

When Mike produced summary statistics he found that the boys scored higher on average than the girls. He could see this in the boxplots he produced too. But when Mike completed an unpaired t-test, he found the results were not statistically significant.

Mike's conclusion will summarise what he did, conclude that there is no statistically significant difference, but will also mention the summary statistics and graphical display he produced. He will say that while there was a difference in the sample, he could not make an inference about the population of all the students in his school.

In order to write a good conclusion, make sure you are able to answer the following questions:

1 What was your research question?

2 What population is your question about?

3 What did you do in your project?

4 Did your hypothesis test suggest a result that was statistically significant or not?

5 What inference can you make about your population?

6 What is the answer to your research question?

7 What descriptive statistics did you generate?

8 Did your descriptive statistics help you answer your question? Why or why not?

9 What graphical displays did you produce?

10 Did your graphical displays help you to answer your question? Why or why not?

11 Double check: have you clearly answered the question you posed in the introduction of your project?

- I have structured my project to include an introduction, subjective impression, analysis and a conclusion.
- I have matched my data type to the correct graphical display and hypothesis test.
- I have concluded my project and I have remembered to refer back to my graphical displays, summary statistics and hypothesis test in my conclusion.
- I have made sure that my project is well presented and contains clear headings and sentences that introduce my graphical displays, summary statistics and hypothesis tests.

11 Financial Products and Inflation

This chapter will show you how to:

- discuss the purpose of different financial products: insurance, savings and credit
- calculate the interest on mortgages, credit cards and other loans
- complete calculations for inflation, savings and credit and analyse the results.

You should already know:

- how to calculate compound interest
- how to reverse a percentage change.

Insurance

The purpose of **insurance** products is to give the **policy holder** protection against a potential **loss**. Policy holders pay an insurance **premium** to the insurer. In the event of an insured **event** occurring, the policy holder can make a **claim**. The amount of money claimed is the amount required to return the policy holder to the same financial position they were in prior to suffering the financial **loss**.

When deciding whether to take out an insurance product, you should consider the following questions:

Question	Example
Does the insurance product cover the **events** you wish it to cover?	If a home insurance policy does not cover flooding, this would be relevant if you lived in an area prone to flooding.
Is there a limit on how much can be claimed, and is a loss likely to be more than this?	You need to be aware of **underinsuring**. If you insure your house for £300 000, this would be too little if your house was worth £500 000.
What would happen if you weren't insured and a loss occurred?	You might decide to not insure your mobile phone if you think you would be able to replace it easily in the event of it being stolen. For some insurance products there is no choice: you must have motor insurance if you intend to drive a car, for example.
Can you afford to pay the premiums?	The insurance must be affordable. You should also consider the length of the term of the policy.
What **exclusions** appear on the policy?	A car insurer may say it won't honour a claim if the car is being driven on a racing track. A home insurer may not cover flooding for houses built in an area susceptible to flooding.

You can take out insurance to cover many different losses. The table below details some common insurance products:

Product	Details
Buildings insurance	Covers damage to a building through things like fires, earthquakes, and burst pipes.
Contents insurance	Covers the items inside your property.
Life insurance	Pays out in the event of the death of the policy holder. When a couple purchase a house with a mortgage, it is common to take out life insurance so that the mortgage can still be paid in the event of the death of one of the partners.
Car insurance	It is a legal requirement in the UK to have 'third party' car insurance. This covers your liability to others in the event of a car accident. For example, if you crash your car into somebody's house, your third party car insurance would pay the cost of repairing the house. Other types of car insurance include 'Fire and Theft' policies, and 'comprehensive' policies which cover damage to one's own vehicle.
Pet insurance	Covers the cost of treating a pet in the event of it becoming injured or ill.
Travel insurance	Covers the cost of becoming injured or ill whilst on holiday, or having to cancel a trip due to injury or illness. Many policies also cover loss of possessions whilst on holiday.

Example 11.1

Michael has purchased his house for £240 000 using a mortgage. He is considering taking out a life insurance policy, some key points of the policy are:

- Value of claim in the event of death: £250 000.
- Coverage: Covered in the event of death (where the policy holder dies within the UK).
- Exclusions: Deaths caused while participating in extreme sports.
- Annual premium: £500.

Give three reasons why Michael may decide not to take out the life insurance.

Michael may wish to take out a policy that he pays monthly, rather than annually.

If Michael travels abroad a lot, or participates in extreme sports, this life insurance might not meet his needs. Michael should check what constitutes an extreme sport, as some insurers might consider a broad range of sports to be 'extreme sports'.

The value of claim may be too high. As Michael is paying off his mortgage the amount owed will reduce over time. Michael may be better off purchasing a cheaper policy whereby the value of the claim reduces over time.

Credit cards and loans

Credit is when you borrow money. For most people, credit is essential to make large purchases, such as a house or a car. Examples of credit products include **credit cards, loans, bank overdrafts** and **mortgages**.

Borrowers must pay back the amount borrowed (the **capital**) and also **interest** on the loan.

Loans are important for being able to afford large purchases, for example when buying a car or a house. A loan allows a person to have money upfront and pay it back in instalments.

Example 11.2

Lucy takes out a personal loan of £4000 for 2 years, with an effective rate of interest of 0.25% per month. She has two options for repaying the loan:

- Option 1: Pay the capital and the interest at the end of the 2-year term in a lump sum.
- Option 2: Make interest payments each month and pay back the capital at the end of the 2-year term.

  a   For each option work out the total interest Lucy would have to pay.

Lucy decides to take Option 2.

  b   Give one reason why Lucy might decide to take Option 2.

a   Option 1

$$£4000 \times 1.0025^{24} = £4247.03$$

> The amount borrowed is appreciated by the monthly interest rate to the power of 24 as there are 24 months in 2 years.

For Option 1 the total repayment amount is £4247.03. The total interest paid by Lucy is £247.03

Option 2

To find the amount of interest to be paid each month, multiply the amount borrowed by the interest rate:

$$£4000 \times 0.0025 = £10$$

This means that Lucy would pay the lender £10 a month, and then pay back the £4000 at the end of the term of the loan. The total interest to be paid is:

$$£10 \times 24 = £240$$

b   Option 2 is cheaper by £7.03.

Savings products

Savings products allow customers to deposit money and earn interest. For products that are **easy access**, customers can withdraw their money at any time. Other accounts may require customers to give notice before making a withdrawal, or place limits on how much can be withdrawn.

Money placed into **bonds** can only be accessed after a fixed amount of time. Easy access products are likely to pay less interest than products with more conditions.

Example 11.3

Sophie is saving towards her wedding, which will take place in 3 years' time. She needs to save £7000. Sophie currently has savings of £6150. She needs to decide between two different savings products.

Product One

- A 3-year, fixed-rate bond.
- Pays an effective rate of interest of 4.5% per annum.
- Interest is paid at the end of the term of the bond.
- Deposit cannot be accessed until the end of the term.

Product Two

- Easy access savings account.
- Variable interest rate, guaranteed rate of 4.5% per annum for the first year.
- Interest is paid annually.
- Deposits can be withdrawn at any time.

   a   Give two risks associated with saving Product One.

   b   Give one reason why Sophie may decide to choose Product One.

a Sophie will not have access to her money before the end of the 3 years. If Sophie needed her money early, for example if she had a sudden unexpected cost to pay then she might be unable to pay.
Because Product One is a **fixed-rate**, Sophie will not benefit if interest rates were to increase. If Sophie chose Product One, she would have £6150 × 1.045^3 = £7018.17 for the wedding. If the cost of the wedding increased then she might not have enough.

b The interest rate for Product Two is **variable**, which means that after 1 year the rate could decrease. This puts Sophie at a considerable risk of not having enough money after 3 years. For example, consider what would happen if the interest rate dropped to 4% after 1 year. £6150 × 1.045 × 1.04^2 = £6951.17. Product One will allow Sophie to be confident that she will meet the £7000 goal.

Exercise 11A

1 Freya takes out a loan for £5000. Here are some details of the loan:

- Interest is charged monthly, at an effective rate of interest of 0.33% per month.
- The loan must be repaid at the end of 3 years.
- Interest can either be paid monthly, or paid back with the capital at the end of the term of the loan.

   a Calculate the total amount of **interest** Freya would pay if she decided to pay interest monthly.

   b Calculate the total amount of **interest** Freya would pay if she delayed paying her interest payments back until the end of the term of the loan.

   c Give two reasons why Freya might decide to pay interest back monthly.

★ 2 Ruairi is considering depositing £400 into a 4-year, fixed-rate savings bond. The bond pays 5% as an annual, effective rate of interest.

   a Calculate the total **interest** he would earn if he deposited the £400 into the bond.

   Ruairi also has a loan for £3000, due to be repaid in 4 years. The interest on the loan is fixed at an effective monthly rate of 0.4%. Ruairi can either make interest payments monthly, or pay interest at the end of the loan.

   b For each option work out the total interest Ruairi would have to pay on the loan.

   Ruairi needs instant access to his £400 savings if he wishes to make interest payments monthly.

   c Give a reason why Ruairi might decide to deposit his £400 into the savings bond rather than make monthly interest payments.

★ 3 Darcy is considering buying home insurance. Darcy bought her home 5 years ago for £125 000. Some key points of the policy are:

   • Type of policy: Buildings and contents cover.
   • Total value insured: £130 000.
   • Term: 3 years.
   • Coverage: The policy will pay out, up to the total value insured, in the event of theft, natural disaster, fire.
   • Exclusions: Accidental damage.

   Give three reasons why Darcy may decide not to buy this policy.

  4 Desmond is considering opening a Regular Savers Account with Lang Toun Building Society. Some key points about the account are:

   • Savers must deposit between £50 and £500 a month.
   • The interest rate is fixed at 2.3% per annum, paid at the end of the term.
   • The account is a bond with a 2-year term. At the end of the term the money in the account can either be withdrawn or re-invested.
   • If money is withdrawn before the end of the term, the interest rate reduces to 0.35% per annum.
   • The maximum investment is £7200.

   Give three reasons why Desmond might decide to not open the account.

★ 5 Zahar is travelling to France on a skiing holiday. Zahar has a current account which offers him free travel insurance. Some key points about this travel insurance are:

   • Type of cover: Emergency medical cover, repatriation in the event of injury or illness, personal belongings.
   • Limits: £1 million for medical cover and repatriation. £250 for personal belongings.
   • Term: Covers up to three trips overseas each calendar year. Each trip must be less than 2 weeks to be covered.

- Exclusions: Injuries caused from excessive alcohol intake, or participation in extreme sports. Illnesses caused by a pre-existing health condition.

Zahar decides to purchase travel insurance from a different insurer, rather than use his free travel insurance. Give three reasons why Zahar might have made this decision.

6 A high street bank offers Maya a choice of three possible loans:

Loan	Term (years)	Monthly interest payments	Monthly effective rate of interest
1	4	Interest payments paid monthly	0.25%
2	2	Interest paid back in one payment at the end of the term of the loan	0.3%
3	3	Interest payments paid monthly	Year 1: 0.25% Year 2: 0.3% Year 3: 0.31%

Maya is borrowing £6000. Calculate the total amount in interest Maya will pay for each loan.

7 Below is part of a car insurance policy document:

Losses not covered

Radioactivity – We won't cover any loss caused, or contributed to, by radioactive, toxic, explosive or other dangerous properties of nuclear equipment or its nuclear parts.

War – We won't cover any loss or injury caused by war, invasion, revolution or a similar event, unless strictly required by the Road Traffic Act.

Why would an insurer want to include these exceptions in its insurance policies?

8 Scott is considering purchasing £6000 in children's bonds for his daughter. The bonds pay an effective rate of interest of 4% per annum, and mature when the child turns 18. The money invested is not accessible until the child turns 18. Scott is considering whether to purchase the bonds when his daughter is born, or purchasing them on her 5th birthday.

How much more would the bonds be worth on maturity if Scott purchased the bonds at birth, rather than at age 5?

9 Rodney owns some golf clubs worth £7000. He stores his golf clubs in his house and transports them to his golf club in his car. Below are some key details of Rodney's home and car insurance.

Home

- Covers personal belongings from theft and damage.

- Includes personal possessions cover for items taken out of the home.

- The policy has a limit of £10 000 for personal possession cover.

Car

- A comprehensive policy that covers damage to third parties, and damage to the policy holder's vehicle.

- Includes cover of items within the policy holder's vehicle.

- Limits: No limit to the claims for third party's losses. The policy will pay to fix damage to the policy holder's vehicle, or in the event the damage is too much, will pay the current market value of the car. Items in the vehicle are covered up to a total value of £500.

Rodney is involved in a road traffic accident, which causes extensive damage to the rear of his car and damages his golf clubs.

Explain why Rodney may need to make a claim through both insurance products.

★ 10 Alan is saving for retirement, which he is planning to take in 10 years. To supplement his pension, Alan wants to have £25 000 in savings when he retires. Alan currently has £16 000. He needs to decide between three different savings products.

A 10-year government bond

- A 10-year, fixed-rate bond.
- Pays an effective rate of interest of 4.75% per annum.
- Interest is paid at the end of the term of the bond.
- Deposit cannot be accessed until the end of the term.

Variable savings account

- Building society savings account.
- Variable interest rate, guaranteed rate of 3.25% per annum for the first 3 years.
- Interest is paid annually.
- Deposits can be withdrawn at any time; however, if a withdrawal is made the interest rate is reduced to 0.05% per annum for that financial year.

Premium bonds

- Bonds that can be easily converted back to cash.
- Premium Bonds don't earn interest – instead you are entered into a monthly prize draw where you can win between £25 and £1 million.
- The probability of winning a prize is 34 500 to 1 (for every £1 Bond).

a   Explain why premium bonds are not a suitable savings product for Alan.

b   Give two risks associated with the variable savings account.

c   Is the 10-year government bond suitable for Alan's needs? Justify your answer with a calculation.

Inflation

Over time the amount of money it costs to purchase a product or service changes. This is called **inflation**. Inflation is the measure of the rate of rising prices. If prices decrease over time, this is called **deflation**. One measure of inflation is the Consumer Price Index (CPI). A way to think about the CPI is to imagine a shopping basket filled with products and services that people typically spend their money on. The CPI looks at how the prices of the items in this basket change over time, and calculate the rate of inflation. The shopping basket of items is regularly updated to reflect current spending patterns.

Inflation is often given as an annual figure.

Example 11.4

The table below shows the price of a Big Mac at different points in time.

Date	Price
2008	£2.29
2015	£2.89
2022	£3.59

Calculate the effective rate of inflation for Big Macs between 2008 and 2015, and between 2015 and 2022.

2008 to 2015: $\frac{£2.89}{£2.29} = 1.262$. That is an inflation rate of 26.2%.

2015 to 2022: $\frac{£3.59}{£2.89} = 1.242$. An inflation rate of 24.2%

> You divide the new price by the old price, then subtract 1.

Example 11.5

Emily estimates that the effective rate of inflation will be 2.5% per year for the next 5 years.

a   Use this estimate to calculate how much a £600 mobile phone will cost in 3 years' time.

b   Emily estimates that a Big Mac will cost £4.50 in 5 years' time. How much is a Big Mac today?

(continued)

c   Emily plans to purchase a new computer in 5 years' time. The computer currently costs £1250. She has £1000 to put into a savings account which pays an effective monthly rate of 0.61%. Assuming the price of the computer will rise with inflation, will Emily have enough money in her account to purchase the computer after 5 years?

d   Describe two risks that could result in Emily not having enough money for her new computer.

a $£600 \times 1.025^3 = £646.13$

b $£4.50 \div 1.025^5 = £3.98$

c   The price of the computer in 5 years would be $£1250 \times 1.025^5 = £1414.26$. She would accumulate in her bank $£1000 \times 1.0061^{60} = £1440.35$.

   Yes, she would have enough money.

   > There are 60 months in 5 years.

d   Emily's savings account interest rate may change, or Emily's assumption about the rate of inflation could be incorrect. It might also be the case that the price of the computer rises faster than the general rate of inflation.

Example 11.6

In 2015, 'Pick and Mix' in a sweet shop costs 55p per 100 g of sweets; 5 years later, it costs 85p per 100 g.

a   Calculate the effective rate of inflation for this period.

b   How many fewer grams of sweets would £5 buy in 2020 compared with 2015?

a   85 ppg ÷ 55 ppg = 1.55, a rate of 55%.

b   In 2015, £5 would buy 909 g of sweets. In 2020, £5 would buy 588 g. This means £5 buys 321 fewer grams in 2020.

> How far our money goes is known as **purchasing power**. With inflation, the amount of purchasing power money has decreased over time.

Exercise 11B

1   In 1994, Cadbury's sold a small chocolate bar, Freddo, for 10 pence. In 2020, Freddos were sold for 25 pence each. Calculate the effective rate of inflation for Freddos over this period.

2   The consumer price index, or CPI, had a value of 48.2 in 1988. In 2005, its value was 86.2. Calculate the effective rate of inflation for this period.

3   The estimated rate of inflation for the next 3 years is 3% per year.

   a   Find the estimated value of a product that costs £40 today, in 3 years' time.

   b   If a product costs £40 000 in 3 years' time, estimate its value today.

★ 4 The table here shows historic data of the CPI:

Date	CPI value
1988	48.2
1989	51.0
1990	55.1
1991	59.2
1992	61.9
1993	63.5
1994	64.9
1995	66.6
1996	68.5
1997	70.0
1998	71.3

a Calculate the effective rate of inflation between 1989 and 1990.

b Calculate the effective rate of inflation between 1990 and 1998.

c Use your answer to part **b** to estimate the cost of a packet of crisps in 1998, when they cost 25p in 1990.

d Use your answer to part **b** to estimate the cost of a fizzy drink in 1990, when the drink cost 50p in 1998.

5 In 2006, petrol cost 89.88 pence per litre. In 2021, the price was 145.99 pence per litre.

a Calculate the effective rate of inflation.

b How many litres of petrol could be purchased for £50 in 2006?

c How many litres of petrol could be purchased for £50 in 2021?

d How many more litres would £50 buy you in 2006, compared with 2021?

6 Billy puts £300 into a savings account in 2023. The effective rate of interest on the savings account is 1.25% per annum. Billy leaves her money in the account for 5 years. Over this period, the annual rate of inflation was 5%.

a Has Billy's savings' purchasing power increased, or decreased?

b Calculate the amount in Billy's account after the 5 years.

c A mobile phone cost £300 in 2023. Assuming the price rose with inflation, how much would it cost after 5 years?

7 Below are some historical petrol prices:

Year	Pence per litre
1983	35.6
1984	38.7
1985	43.8
1986	38.3
1987	37.8
1988	35.7
1989	38.4
1990	40.2
1991	39.5
1992	40.3
1993	44.9
1994	48.9
1995	50.8
1996	52.9
1997	57.9
1998	60.9

a Calculate the effective rate of inflation between 1983 and 1998 for petrol.

b How many more litres of petrol would £100 buy in 1991 compared with 1990?

c Lily's car can hold 42.15 litres of petrol. How much money would it take to fill her car in 1983? Give your answer to the nearest pound.

d Lily decides to always put £15 of fuel into her car. What percentage of her tank is filled when she does this in 1987?

8 Sandra is making plans for retirement. She estimates that when she retires, she will need £1200 a month. She estimates that this amount will increase due to inflation by 1.5% each year.

a How much money will she need per month after 5 years of retirement?

b Sandra buys a new car for £32 000 when she is 8 years into retirement. Assuming the cost of the car has risen with inflation, calculate how much the car would have cost if she had bought it when she first retired. ***Give your answer to 3 significant figures***.

★ 9 Andrea receives a gift of £1000 on her 25th birthday. She is deciding whether to spend all the money on a holiday this year or save the money and spend it on a holiday when she turns 30. Andrea has access to a savings account that pays an effective monthly rate of interest of 0.25%. She assumes that the cost of the holiday will rise with inflation of 2.1% per year.

a When will Andrea have a greater amount of purchasing power for the holiday, now or when she is 30? Justify your answer with calculations.

b Describe two risks that could result in Andrea not having enough money to go on her chosen holiday when she turns 30.

⊕★ **10** Open the spreadsheet file named 11B.

    a Michael plans to retire when he is 67. He estimates that his annual cost of living will be £15 000, and that will increase by 2.5% per year due to inflation. Complete the spreadsheet "A" to show how much Michael will need per year after he retires.

    b The second tab has some historical petrol prices. Use these to complete the sheet "B". Find the amount of litres of petrol that £50 would buy in each year, and then express this as a percentage of the 2010 amount.

    c In tab "C", calculate the effective annual rate of inflation between 2010 and each subsequent year.

- I can discuss the purpose of different financial products: insurance, savings and credit. ★ Exercise 11A Q3, 5
- I can calculate the interest on loans such as mortgages, credit cards and other loans. ★ Exercise 11A Q2, 10
- I can complete calculations for inflation, savings and credit and analyse the results. ★ Exercise 11B Q4, 9, 10

12 Accumulation and Present Value (Calculator)

This chapter will show you how to:

- calculate an accumulated value
- express effective rates of interest in different time frequencies
- calculate the present value
- calculate accumulated and present values with varying interest rates
- calculate accumulated and present values with multiple payments.

You should already know:

- how to calculate compound interest
- how to reverse a percentage change.

Accumulated value

Money can be placed into savings products, such as savings accounts or bonds. The amount of money deposited is called **capital**. Savings products pay **interest**. For example, if you placed £400 into a savings account last year you might find it is worth £430 today. The **capital** was £400, the **interest** earnt is £30.

Accumulated value is the value that the capital has grown to, at a particular time. You would say that your £400 capital now has an **accumulated value** of £430.

In simple situations, a single payment is made into a savings product and accumulates over time. Later in the chapter you will consider the accumulation of multiple payments.

Example 12.1

Oliver deposits £70 into a savings account. The effective rate of interest is 2% per annum. Find the accumulated value after:

a   3 years.
b   6 months.
c   5 years and 3 months.

To accumulate a value, you multiply the capital by the percentage multiplier to the power of the time period. In this case, as the interest is paid annually, the index must be in years.

a £70 × 1.02^3 = £74.28 ●————————

b £70 × 1.02$^{0.5}$ = £70.70 ●

c £70 × 1.02$^{5.25}$ = £77.67

> An increase in 2% is represented as 1.02. The index is 3 as you are accumulating for 3 years.

> 6 months in years is 0.5.

Time frequencies

The **effective rate of interest** is the percentage rate of interest for a specific period of time. An effective rate of 4% per year means that for every £1 you invest, you will be paid 4 pence after 1 year.

You often need to convert interest rates from one unit of time to another, for example if you want to know what a monthly rate is as an annual effective rate. It is not enough to simply multiply the rate by 12.

To convert between time frequencies follow these steps:

1 Write the interest rate as a decimal multiplier.

2 Adjust the index of the multiplier. The index will be how you convert from the new time frequency to the old. For example, to convert from months to years, multiply by 12. When converting from years to months, you would use $\frac{1}{12}$.

3 To convert your answer back to a percentage, subtract one.

This can be expressed with the following formula:

New rate = $(1 + i)^{\frac{y}{x}} - 1$

where i is the interest rate over x time frequency, and y is the new time frequency.

A monthly interest rate of 0.5% can be converted to an annual rate:

$(1 + 0.005)^{\frac{12}{1}} - 1 \approx 0.062$

so is equivalent to an annual effective rate of around 6.2%.

Example 12.2

Express these interest rates in different time frequencies:

a What is 10% per year as a monthly interest rate?

b What is 0.3% per month as an annual interest rate?

c What is 0.88% per quarter as a monthly interest rate?

(continued)

> **a** Increasing by 10% is represented by multiplying by 1.1. A month is $\frac{1}{12}$ of a year. $1.1^{\frac{1}{12}} = 1.00797$. This is an interest rate of 0.797% per month.
>
> **b** The multiplier is 1.003. A year is 12 months. $1.003^{12} = 1.0366$. The interest rate is 3.66% per year.
>
> **c** The multiplier is 1.0088. There are 3 months in a quarter, so $\frac{1}{3}$. $1.0088^{\frac{1}{3}} = 1.0029$, which means the interest rate is 0.29% per month.
>
> You might be tempted to think that a monthly interest rate can be found by dividing by 12, but this would not take into account the effect of compounding interest.

Exercise 12A

1 Stuart deposits £450 into his savings account which pays 4% per annum. Calculate how much he will have in his account after:

    a 4 years

    b 7 months

    c 3 quarters

    d 7 quarters

    e 6 years and 4 months.

★ 2 Ella wins £125 from her premium bonds. She puts the money into a savings account with an interest rate of 5% per annum. How much interest will she be due after 3 years and 9 months?

3 Convert the following annual interest rates to monthly rates:

    a 10% p.a.

    b 3% p.a.

    c 5.1% p.a.

    d 0.4% p.a.

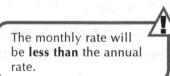

The monthly rate will be **less than** the annual rate.

★ 4 Convert the following interest rates:

    a What is 0.02% per month as an annual interest rate?

    b What is 3% per quarter as an annual interest rate?

    c What is 7% per half year as an annual rate?

5 Amber has £300 to put into a bank account. She plans to save it for 4 years and 6 months. Kirkcaldy Bank offers an interest rate of 4.45% per annum. Kingdom Bank offers an interest rate of 2.3% per half year. Which bank would give Amber the best return?

6 Cassey is deciding between a savings account that pays 5% interest per half year, and another account that pays 10% per year. Cassey says: "The two accounts have the same effective rate." Show that Cassey is incorrect.

★ 7 Sharon wishes to place £540 into a credit union account for 8 months. For each account given here, calculate the amount of money she would have after the 8 months. Which account should she choose?

   a Star Saver: 2.3% per annum, comes with free travel insurance.

   b Happy Saver: 0.57% per quarter, comes with a free Parker pen.

   c Extreme Saver: 0.3797% every 2 months, comes with an information leaflet.

8 Calculate the accumulated value of:

   a £1740 at 0.5% per quarter for 2 months.

   b £1250 at 0.3% per month for 6 years and 8 months.

   c £40000 at 2% per year for 171 days.

Present value

Present value is the amount of capital required to reach a particular goal in the future. Present value takes into account interest paid. Imagine that you want to buy some solar panels in 18 months' time. The panels cost £10000. How much money would you need to put into an interest-paying savings account, today, to reach that target?

Finding the present value is similar to reversing a percentage change. The process is similar to finding an accumulated value except you will use **division**. **Discounting** is another word for finding the present value.

Example 12.3

Michael needs £5000 for a holiday he is taking in 15 months' time. His savings account pays interest at 4.5% per annum. How much must he invest today to meet his goal?

$£5000 \div 1.045^{\frac{15}{12}} = £4732.33$

> Reversing a 4.5% increase is done by dividing by 1.045. The index is the 15 months divided by the 12 months in a year.

Exercise 12B

1 Faron sets a saving goal of £10000 for a new car. She will invest money in a savings bond that lasts 5 years and pays 4% interest per year. What is the minimum amount she needs to invest in the bond?

2 Georgia needs to save up £2500 for a holiday. She places a one-off amount of money in a 2-year bond that pays an effective rate of 5% annually. How much does she need to place in the bond to have enough for her holiday when the bond matures?

★ 3 Teddy wants to go inter-railing around Europe when he is 18. He estimates that he will need £4000. His savings account pays 0.5% interest per quarter. Teddy wonders whether he will have enough money if he places all of the money he receives for his 15th birthday into the account. How much money would he need to get on his 15th birthday to have accumulated enough by his 18th?

4 Callum needs £2500 in 18 months. His savings account pays interest at 2.2% per half-year. How much must he invest today to meet his goal?

5 Tamara is told she will be paid £4000 in 2 years' time. Tamara has access to a savings account that pays 0.3% interest per month. Find the present value of the £4000 payment.

Variations

So far you have worked out the present and accumulated values of single payments. You will now consider multiple payments and varying interest rates.

Example 12.4

Megan's bank account has the following effective rate of interest:

- 5% per year between 1 January 2022 and 1 January 2023.
- 3.25% per half year between 1 January 2023 and 1 January 2024.
- 0.2% per month between 1 January 2024 and 1 January 2025.

Megan deposits £3000 into her account on 1 January 2022. Calculate the accumulated value in her account on 1 January 2025.

This can be completed in just one calculation:

$$£3000 \times 1.05 \times 1.0325^2 \times 1.002^{12} = £3439.56$$

This calculation is just a compact way of doing:

Year 1: £3000 × 1.05 = £3150

Year 2: £3150 × 1.0325^2 = £3358.08 ← Squared as there are two half years in a year.

Year 3: £3358.08 × 1.002^{12} = £3439.56 ← 12 for the 12 months in a year.

Megan would have £3439.56 in her account on the 1 January 2025.

Example 12.5

Hannah's bank account has an effective rate of interest of 4.1% per year. The following payments are made in and out of her bank account:

- £750 paid in after 1 year.
- £400 paid out after 3 years.
- £560 paid in after 6 years.

Calculate:

a The accumulated value after 7 years.

b The present value at year zero.

c The present value after 4 years.

It often makes sense to draw a timeline for more complex questions like these:

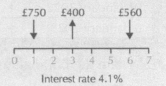

a The calculation you need is:

$£750 \times 1.041^6 - £400 \times 1.041^4 + £560 \times 1.041^1 =$ £1067.69.

The £750 was in the account earning interest for 6 years. The £400 was paid out, so remove (subtract) the £400 and the interest it would have earnt in the remaining 4 years. Add in the £560 and the interest it earnt for 1 year.

b As you know the accumulated value at year 7, you can just ask: "How much would you need to have at year zero (that is, now) to accumulate £1067.69 in 7 years?". This calculation is: $£1067.69 \div 1.041^7 = £805.92$.

Another way to find the present value is to consider the present value of each payment and sum them:

$(£750 \div 1.041^1) + (-£400 \div 1.041^3) + (£560 \div 1.041^6) = £805.92$.

c To find the present value after 4 years there are lots of different options.

> Each term features a division because year zero is **before** all the transactions occur. The index is the number of years after year zero the transaction occurred.

You could start with the value at 7 years (£1067.69) and then discount 3 years:

$£1067.69 \div 1.041^3$

You could start with year zero (£805.92) and accumulate 4 years:

$£805.92. \times 1.041^4$

> Notice that for year 4, you must accumulate (×) the transactions that occurred **before**, and discount (÷) the transaction that occurred **after**.

Alternatively, you could consider each transaction in turn:

$£750 \times 1.041^3 - £400 \times 1.041^1 + £560 \div 1.041^2$

All three of these calculations return an answer of £946.44.

It is important to understand that you are not finding the amount in Hannah's bank on the 4th year. The **present value** takes into account the fact that Hannah has another £560 coming to her in the future.

Example 12.6

Scott has a savings account. The effective rates of interest were as follows:

- 5% per year during 2020
- 2% per half year from 1 January 2021 until 1 July 2022
- 1% per month from 1 July 2022.

(continued)

Scott makes the following transactions, detailed in the table:

Date	Transaction
01 July 2020	£500 deposit
01 January 2021	£300 deposit
01 January 2022	£500 deposit
01 January 2023	£450 withdrawal

Calculate the balance of Scott's savings account on 1 January 2024.

A timeline makes this question much easier to think about:

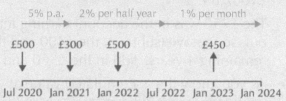

Method 1 – Accumulate each transaction

Using the timeline you can see that:

- The first £500 earns 5% interest per year for half a year, then 2% per half year for one and a half years and then 1% per month for 18 months.
- The £300 earns 2% interest per half year for one and a half years, then earns 1% per month for 18 months.
- The second £500 earns 2% per half year for 1 half year, then earns 1% per month for 18 months.
- Finally, £450 is withdrawn and so misses out on 12 months of 1% per month interest.

  Expressed as a calculation this is:

  $£500 \times 1.05^{0.5} \times 1.02^3 \times 1.01^{18} + £300 \times 1.02^3 \times 1.01^{18} + £500 \times 1.02^1 \times 1.01^{18} - £450 \times 1.01^{12} = £1134.13$

Method 2 – Break down into time periods

Using the same timeline you can think about important dates and work out the accumulated value at each date.

Date	Calculation	Accumulated value
1 July 2020	Initial deposit	£500
1 January 2021	$£500 \times 1.05^{0.5} + £300$	£812.35
1 January 2022	$£812.35 \times 1.02^3 + £500$	£1345.17
1 July 2022	$£1345.17 \times 1.02^1$	£1372.07
1 January 2023	$£1372.07 \times 1.01^6 - £450$	£1006.48
1 January 2024	$£1006.48 \times 1.01^{12}$	£1134.13

Make a new row in your table for every date where either a transaction occurs, or the interest rate changes. Always draw a timeline which clearly shows the interest rates and dates.

Exercise 12C

1 Suppose that the effective rates of interest on your bank account are:

- 4% per year between 1 January 2019 and 1 January 2020
- 3.15% per half year between 1 January 2020 and 1 January 2021
- 0.1% per month between 1 January 2021 and 1 January 2022.

If you deposit £300 on 1 June 2019, calculate its accumulated value on 1 April 2021.

★ 2 Iona took out a loan of £5000 on 1 August 2020. The interest rates for the loan are as follows:

Dates	Interest rates (annual)
1 August 2020 to 31 July 2021	4%
1 August 2021 to 31 November 2021	5.5%
1 December 2022 to 31 March 2023	6%

Assuming Iona does not make any payments in the interim, how much will she owe on 1 April 2023?

3 Rebecca invested £150 into the S&P 500 index on 1 January 2018. She sold her stock on 1 January 2022. The effective rate of return for each year is displayed in the table.

Year	Return (annual)
2021	17.14%
2020	16.26%
2019	2.888%
2018	−6.24%

a Use the information in the table to calculate how much money she made in this time.

b What was the effective rate of interest for this time period all together?

★ 4 Reece will receive a payment of £4000 in two years as part of his contract. There was an effective interest rate of 2.5% per annum available to him.

a Find the present value of this payment.

b Reece renegotiates his contract. He will now receive a payment of £2000 in one year, and £2100 in three years. Find the new present value.

5 A life insurer gives a claimant two options. Option A: take a lump sum of £50 000 immediately, or Option B: take £10 500 next year and then annually for a further 4 years (so 5 payments in total).

a Assuming an effective interest rate of 3% is available for those 5 years, find the present value of Option B.

b The life insurer has a third option, Option C: receive a pay-out of £56 885 in 5 years' time. Place the three options in order of accumulated value after 5 years.

6 Sophie has a bank account with £0 in it on 1 January 2021. The interest rate for the account is 3%. Over the course of the next 6 years, Sophie makes the following transactions, always on the 1 January:

Year	In	Out
2022	£300	–
2024	–	£50
2026	£450	–

a Find the accumulated value on 1 January 2027.

b Find the present value on 1 January 2025.

c Find the present value on 1 January 2021.

d Explain why your answer to part **c** is not zero.

7 Cameron's bank account has an effective rate of interest of 5% per year. The following payments are made in and out of his bank account:

- £9914.82 paid in after 1 year
- £200 paid out after 2 years
- £500 paid in after 4 years.

Calculate:

a The accumulated value after 5 years.

b The present value at year 0.

c The present value after 3 years.

★ 8 Lauren sets herself a savings challenge: She will put £10 in her savings account at the **end** of January, £20 at the end of February, £40 at the end of March, and so on, doubling each month until the end of December. Lauren's bank account offers an effective rate of interest of 5% per year. Calculate the balance of Lauren's saving account at the end of the year, after the payment made on 31 December.

★ 9 Harry started university on 1 September 2018 and finished on 30 June 2022. He received a student loan of £4000 on 1 September each year while he studied. The student loan interest rates were:

Date	Interest rate (annual)
1 September 2018 to 31 August 2019	1.25%
1 September 2019 to 31 August 2020	1.5%
1 September 2020 to 31 January 2022	2%
1 February 2022 to 31 August 2022	4%

How much did Harry owe in student loans when he finished university?

10 Bernoulli is playing around with compound interest. He writes down the following questions. How many times bigger is the accumulated value than the original capital for each case?

 a £1 at 100% per year for 1 year.

 b £1 at 50% per half year for 1 year.

 c £1 at 25% per quarter for 1 year.

 d £1 at $\frac{100}{12}$% per month for 1 year.

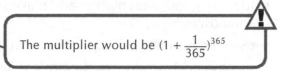

The multiplier would be $(1 + \frac{1}{12})^{12}$

 e £1 at $\frac{100}{365}$% per day for 1 year.

The multiplier would be $(1 + \frac{1}{365})^{365}$

 f £1 at $(100 \div 31\,536\,000)$% per second for one year.

As the time period gets shorter and shorter, what happens to the ratio of accumulated value to original capital?

- I can calculate an accumulated value. ★ Exercise 12A Q2, 7
- I can express effective rates of interest in different time frequencies. ★ Exercise 12A Q4
- I can calculate a present value. ★ Exercise 12B Q3
- I can calculate accumulated and present values with varying interest rates. ★ Exercise 12C Q2, 9
- I can calculate accumulated and present values with multiple payments. ★ Exercise 12C Q4, 8, 9

13 Accumulation and Present Value with Spreadsheets

This chapter will show you how to:

- use spreadsheets to carry out calculations in a way that another independent user can easily understand
- identify good presentation and make suggestions to improve poorly presented spreadsheets
- use spreadsheets to find the accumulated value of a series of payments
- use spreadsheets to find the present value of a series of payments.

You should already know how to:

- express effective rates of interest in different time frequencies
- calculate accumulated and present values using a calculator
- implement recurrence relations in a spreadsheet
- use a variety of spreadsheet functions.

Creating spreadsheets that can be understood

It is important that any spreadsheet you create is easy for somebody else to understand. This ensures that mistakes are easy to spot, and that the information you wish to convey is easily picked up by the reader. To make your spreadsheet readable you should:

- include a title which describes what your spreadsheet does
- use **key** or **base variables** and give them suitable names
- give columns suitable names
- label formulae.

Key variables are values that you will make **absolute references** to in your spreadsheet. Common examples of key variables are interest rates, regular payment amounts and initial deposits. Key variables are useful as they allow you to update your spreadsheet by changing only one cell. Example 13.1 will show you how to set up key variables and make absolute references to them.

> In the exam you will need to print your spreadsheet in two formats: regular view and formula view. To change between views, use the keyboard shortcut CTRL+`. The ` key is usually located near the TAB key.

Example 13.1

Patryk plans on depositing money into his savings account at the **end** of every month. His savings account has an effective rate of interest of 5% per annum. In January, Patryk will deposit £50. Each month, Patryk will deposit 10% more than the previous month.

a How much will Patryk have in his account on 31 December?

b How much interest will he earn over the year?

You can open the spreadsheet called "13.1".

Student 1

Below is student 1's answer in both regular and formula view:

	A	B	C	D
1	£50	11	£52.29	
2	£55	10	£57.28	
3	£61	9	£62.75	
4	£67	8	£68.75	
5	£73	7	£75.32	
6	£81	6	£82.52	
7	£89	5	£90.40	
8	£97	4	£99.04	
9	£107	3	£108.50	
10	£118	2	£118.86	
11	£130	1	£130.22	
12	£143	0	£142.66	
13			£1,088.59	£19.35

	A	B	C	D
1	50	11	=ROUND(A1*1.05^(B1/12),2)	
2	=ROUND(A1*1.1,2)	10	=ROUND(A2*1.05^(B2/12),2)	
3	=ROUND(A2*1.1,2)	9	=ROUND(A3*1.05^(B3/12),2)	
4	=ROUND(A3*1.1,2)	8	=ROUND(A4*1.05^(B4/12),2)	
5	=ROUND(A4*1.1,2)	7	=ROUND(A5*1.05^(B5/12),2)	
6	=ROUND(A5*1.1,2)	6	=ROUND(A6*1.05^(B6/12),2)	
7	=ROUND(A6*1.1,2)	5	=ROUND(A7*1.05^(B7/12),2)	
8	=ROUND(A7*1.1,2)	4	=ROUND(A8*1.05^(B8/12),2)	
9	=ROUND(A8*1.1,2)	3	=ROUND(A9*1.05^(B9/12),2)	
10	=ROUND(A9*1.1,2)	2	=ROUND(A10*1.05^(B10/12),2)	
11	=ROUND(A10*1.1,2)	1	=ROUND(A11*1.05^(B11/12),2)	
12	=ROUND(A11*1.1,2)	0	=ROUND(A12*1.05^(B12/12),2)	
13			=SUM(C1:C12)	=C13-SUM(A1:A12)

This spreadsheet is difficult to understand. It lacks a title and does not use key variables. The columns are not labelled and it is difficult to know if the formulae are correct.

(*continued*)

Student 2

	A	B	C	D	E	F
1		Accumulated Value on 31 December				
2		Initial Deposit	£50			
3		Percentage Increase each month	10%			
4		Increase multiplier	1.1			
5		Interest rate annual	5%			
6		Interest rate monthly	0.41%			
7						
8		Month	Months Remaining	Amount Deposited	Accumulated Value of Payment on 31 December	Rounded Accumulated Value of Payment on 31 December
9		Jan	11	£50.00	£52.29	£52.29
10		Feb	10	£55.00	£57.28	£57.28
11		Mar	9	£60.50	£62.75	£62.75
12		Apr	8	£66.55	£68.75	£68.75
13		May	7	£73.21	£75.32	£75.32
14		Jun	6	£80.53	£82.52	£82.52
15		Jul	5	£88.58	£90.40	£90.40
16		Aug	4	£97.44	£99.04	£99.04
17		Sep	3	£107.18	£108.50	£108.50
18		Oct	2	£117.90	£118.86	£118.86
19		Nov	1	£129.69	£130.22	£130.22
20		Dec	0	£142.66	£142.66	£142.66
21					Total in account on December 31st:	£1,088.59
22					Total Deposited:	£1,069.24
23					Total Interest:	£19.35
24						

	A	B	C	D	E	F
1				Accumulated Value on 31 December		
2		Initial Deposit	50			
3		Percentage Increase each month	0.1			
4		Increase multiplier	=1+C3			
5		Interest rate annual	0.05			
6		Interest rate monthly	=(1+C5)^(1/12)-1			
7						
8		Month	Months Remaining	Amount Deposited	Accumulated Value of Payment on 31 December	Rounded Accumulated Value of Payment on 31 December
9		Jan	11	=$C$2	=D9*(1+$C$6)^(C9)	=ROUND(E9,2)
10		Feb	10	=ROUND(D9*$C$4,2)	=D10*(1+$C$6)^(C10)	=ROUND(E10,2)
11		Mar	9	=ROUND(D10*$C$4,2)	=D11*(1+$C$6)^(C11)	=ROUND(E11,2)
12		Apr	8	=ROUND(D11*$C$4,2)	=D12*(1+$C$6)^(C12)	=ROUND(E12,2)
13		May	7	=ROUND(D12*$C$4,2)	=D13*(1+$C$6)^(C13)	=ROUND(E13,2)
14		Jun	6	=ROUND(D13*$C$4,2)	=D14*(1+$C$6)^(C14)	=ROUND(E14,2)
15		Jul	5	=ROUND(D14*$C$4,2)	=D15*(1+$C$6)^(C15)	=ROUND(E15,2)
16		Aug	4	=ROUND(D15*$C$4,2)	=D16*(1+$C$6)^(C16)	=ROUND(E16,2)
17		Sep	3	=ROUND(D16*$C$4,2)	=D17*(1+$C$6)^(C17)	=ROUND(E17,2)
18		Oct	2	=ROUND(D17*$C$4,2)	=D18*(1+$C$6)^(C18)	=ROUND(E18,2)
19		Nov	1	=ROUND(D18*$C$4,2)	=D19*(1+$C$6)^(C19)	=ROUND(E19,2)
20		Dec	0	=ROUND(D19*$C$4,2)	=D20*(1+$C$6)^(C20)	=ROUND(E20,2)
21					Total in account on December 31st:	=SUM(F9:F20)
22					Total Deposited:	=SUM(D9:D20)
23					Total Interest:	=F21-F22
24						

This student has given the spreadsheet a title, and clearly labelled the key variables. It is clear to the reader that the student has decided to accumulate each payment, and then find the sum to work out the amount on 31 December.

> Whenever you refer to a key variable, make sure to use an absolute reference using $ signs. Remember you can just press F4.

Cell C6 converts an annual rate to a monthly rate. This is one of the most common formulae you will need to use when creating a spreadsheet. Remember that to convert an interest rate from years to months you must:

- add one to make the interest rate into a decimal multiplier
- raise to the power of $\frac{1}{12}$
- subtract one to convert the multiplier back into an interest rate.

(continued)

The formulae in the E column accumulate each value for the correct number of months. The student has used a different colour in cells F21 and F23 to make the answers stand out. You can see that the interest earnt is the total amount paid into the account subtract the final balance.

Student 3

Accumulated Value on 31 December			
Initial Deposit	£50		
Percentage Increase each month	10%		
Increase multiplier	1.1		
Interest rate annual	5%		
Interest rate monthly	0.41%		
Date	Amount Deposited	Total Balance	Total balance (Rounded)
31-Jan	£50	£50	£50.00
28-Feb	£55.00	£105.20	£105.20
31-Mar	£60.50	£166.13	£166.13
30-Apr	£66.55	£233.36	£233.36
31-May	£73.21	£307.52	£307.52
30-Jun	£80.53	£389.30	£389.30
31-Jul	£88.58	£479.47	£479.47
31-Aug	£97.44	£578.86	£578.86
30-Sep	£107.18	£688.40	£688.40
31-Oct	£117.90	£809.10	£809.10
30-Nov	£129.69	£942.09	£942.09
31-Dec	£142.66	£1,088.59	£1,088.59
Total:	£1,069.24		
Total interest:	£19.35		

	A	B	C	D	E
1			Accumulated Value on 31 December		
2		Initial Deposit	50		
3		Percentage Increase each month	0.1		
4		Increase multiplier	=1+C3		
5		Interest rate annual	0.05		
6		Interest rate monthly	=(1+C5)^(1/12)-1		
7					
8		Date	Amount Deposited	Total Balance	Total balance (Rounded)
9		44592	=C2	=C9	=ROUND(D9,2)
10		44620	=ROUND(C9*$C$4,2)	=E9*(1+$C$6)+C10	=ROUND(D10,2)
11		44651	=ROUND(C10*$C$4,2)	=E10*(1+$C$6)+C11	=ROUND(D11,2)
12		44681	=ROUND(C11*$C$4,2)	=E11*(1+$C$6)+C12	=ROUND(D12,2)
13		44712	=ROUND(C12*$C$4,2)	=E12*(1+$C$6)+C13	=ROUND(D13,2)
14		44742	=ROUND(C13*$C$4,2)	=E13*(1+$C$6)+C14	=ROUND(D14,2)
15		44773	=ROUND(C14*$C$4,2)	=E14*(1+$C$6)+C15	=ROUND(D15,2)
16		44804	=ROUND(C15*$C$4,2)	=E15*(1+$C$6)+C16	=ROUND(D16,2)
17		44834	=ROUND(C16*$C$4,2)	=E16*(1+$C$6)+C17	=ROUND(D17,2)
18		44865	=ROUND(C17*$C$4,2)	=E17*(1+$C$6)+C18	=ROUND(D18,2)
19		44895	=ROUND(C18*$C$4,2)	=E18*(1+$C$6)+C19	=ROUND(D19,2)
20		44926	=ROUND(C19*$C$4,2)	=E19*(1+$C$6)+C20	=ROUND(D20,2)
21		Total:	=SUM(C9:C20)		
22		Total interest:	=E20-C21		
23					

This spreadsheet is also easy to understand. This student has used a different approach, finding the value in the account at the end of each month.

It is important to remember to use the ROUND function when calculating with money. In this example the "Total Balance" and "Total balance (Rounded)" column look identical in regular view. This is because the spreadsheet is only displaying the number to two decimal places. If you adjust the spreadsheet to show numbers to 3 decimal places, you will see the difference:

(continued)

Total Balance	Total balance (Rounded)
£50.000	£50.000
£105.204	£105.200
£166.129	£166.130
£233.357	£233.360
£307.521	£307.520
£389.303	£389.300
£479.466	£479.470
£578.863	£578.860
£688.398	£688.400
£809.105	£809.100
£942.086	£942.090
£1,088.588	£1,088.590

Exercise 13A

Complete this exercise in the spreadsheet file called 13A. The first sheet in this spreadsheet contains some historic currency exchange rates.

1   Convert the annual interest rates of 4%, 2.5% and 11% into monthly effective rates, and quarterly effective rates.

2   Sandra opens a savings account on the first of February with an initial deposit of £500. She will deposit £50 at the beginning of every month. Her savings account has an effective annual interest of 2.1%. Complete the spreadsheet to find her balance at the end of October.

★ 3   Erin plans on depositing money into her savings account at the **end** of every month. Her savings account has an effective rate of interest of 3% per annum. In January Erin will deposit £25. Each month, Erin will deposit 2% more than the month previous. Use a spreadsheet to:

   a   Find the accumulated balance on the 31 December.

   b   Find the total interest earnt over the year.

4   Haley has £1000 in her savings account. Starting in August, she will withdraw money at the beginning of each month. In August, she withdraws £10. Each month she will withdraw £1 more than the month before, so in September she will withdraw £11. Haley's savings account has an effective interest rate of 2.3% per annum.

   a   Create a spreadsheet to find the amount in her account after 1 year.

   b   How many months will pass before Haley has no money remaining in her account?

★ 5   Each year Andrew converts some money from pounds (£) to United States dollars ($) and deposits it into a savings account. Complete the spreadsheet to find the balance, in $, on 1 January 2021. Use the historic exchange from the first sheet of the spreadsheet.

6 Michaela has a savings account with an effective rate of 3% per annum. The table below shows the transactions she made over a year.

Date	In	Out
1 Jan	£700	–
1 March	£400	–
1 July	£250	–
1 August	–	£400
1 October	£100	–

a Create a spreadsheet to calculate the **present value** of these transactions at the start of the year.

b Use a calculator to calculate the **accumulated value** of these transactions on 1 January the following year.

7 Marty is wanting to compare the interest rates between two different savings accounts. He creates a spreadsheet to compare investing £400 for 6 years in two different accounts.

a Marty's spreadsheet can be seen in sheet 7. Give three criticisms of Marty's spreadsheet.

b Old Town Building Society offers a rate of 3% per annum. The New Town ISA offers a rate of 0.25% per month. Improve Marty's spreadsheet to make it better presented.

c Marty wants to know what the difference in interest would be when comparing £1000 deposited in Old Town Building Society for 6 years with the same amount deposited in the New Town ISA for 6 years. Update your spreadsheet to calculate the difference.

Example 13.2

Rylie wishes to save up to buy a car. He plans to put £100 into his bank account at the start of every month, beginning in August 2022. Below are the effective rates of interest for his savings account:

Dates	Annual effective rates of interest (%)
01 April 2022–31 March 2023	3
01 April 2023–31 March 2024	3.5

Find the total amount in Rylie's savings account on 1 December 2023.

(continued)

Open the spreadsheet "13.2".

First you must set up the key variables. In this case, the interest rates and the monthly payments.

	A	B	C	D
1	**Calculation of savings account balance at 31 December 2023**			
2				
3	Description			Value
4	Annual interest rate 01/04/2022 - 31/03/2023			0.03
5	Monthly effective rate 01/04/2022 - 31/03/2023			=(1+D4)^(1/12)-1
6	Annual interest rate 01/04/2023 - 31/03/2024			0.035
7	Monthly effective rate 01/04/2023 - 30/03/2024			=(1+D6)^(1/12)-1
8	Monthly payment			100
9				
10	**Savings account balance at 31 December 2023**			=D30
11				

The "Payment" column and "Balance after payment" column are self-explanatory. To find the account balance at the start if each month, you multiply the previous months' balance by 1 plus the monthly interest rate:

	Time	Account balance	Payment	Balance after
12				
13	(months)	before payment (£)	(£)	payment (£)
14	1	0	=$D$8	=B14+C14
15	2	=ROUND(D14*(1+$D$5), 2)	=$D$8	=B15+C15
16	3	=ROUND(D15*(1+$D$5), 2)	=$D$8	=B16+C16
17	4	=ROUND(D16*(1+$D$5), 2)	=$D$8	=B17+C17
18	5	=ROUND(D17*(1+$D$5), 2)	=$D$8	=B18+C18
19	6	=ROUND(D18*(1+$D$5), 2)	=$D$8	=B19+C19
20	7	=ROUND(D19*(1+$D$5), 2)	=$D$8	=B20+C20
21	8	=ROUND(D20*(1+$D$5), 2)	=$D$8	=B21+C21
22	9	=ROUND(D21*(1+$D$5), 2)	=$D$8	=B22+C22
23	10	=ROUND(D22*(1+$D$7), 2)	=$D$8	=B23+C23
24	11	=ROUND(D23*(1+$D$7), 2)	=$D$8	=B24+C24
25	12	=ROUND(D24*(1+$D$7), 2)	=$D$8	=B25+C25
26	13	=ROUND(D25*(1+$D$7), 2)	=$D$8	=B26+C26
27	14	=ROUND(D26*(1+$D$7), 2)	=$D$8	=B27+C27
28	15	=ROUND(D27*(1+$D$7), 2)	=$D$8	=B28+C28
29	16	=ROUND(D28*(1+$D$7), 2)	=$D$8	=B29+C29
30	17	=ROUND(D29*(1+$D$7), 2)	=$D$8	=B30+C30

The dates on the right-hand side are not required, but do help you to check when to change to the next interest rate. The interest rate changes in April, so the first cell with the new interest rate will be B23.

> Key variables are referred to using absolute references. Interest calculations are rounded to 2 decimal places.

(continued)

Payment (£)	Balance after payment (£)			
£100.00	£100.00	1	August	2022
£100.00	£200.25	1	September	2022
£100.00	£300.74	1	October	2022
£100.00	£401.48	1	November	2022
£100.00	£502.47	1	December	2022
£100.00	£603.71	1	January	2023
£100.00	£705.20	1	February	2023
£100.00	£806.94	1	March	2023
£100.00	£908.93	1	April	2023
£100.00	£1,011.54	1	May	2023
£100.00	£1,114.44	1	June	2023
£100.00	£1,217.64	1	July	2023
£100.00	£1,321.14	1	August	2023
£100.00	£1,424.93	1	September	2023
£100.00	£1,529.02	1	October	2023
£100.00	£1,633.41	1	November	2023
£100.00	£1,738.10	1	December	2023

As you can see, the answer is £1738.10.

Exercise 13B

⬇ Complete this exercise in the spreadsheet file called "13B".

1   Danni is saving towards a holiday. She plans to put £25 into her bank account at the start of every month, beginning in September 2024. The effective rates of interest for her savings account are given in the table:

Year (Jan–Dec)	Annual effective rates of interest (%)
2024	2.6
2025	2.25

How much money will Danni put towards her holiday by 1 June 2025?

2   Craig is saving up a deposit for a house. He already has £30 000 in his savings account. He decides to place an additional £100 into his savings account on 1 December 2023. Each month he will deposit 2% more than the previous month. He needs £40 000 by the 1 January 2027.

The effective rates of interest for his account are given in the table:

Year (Jan–Dec)	Annual effective rates of interest (%)
2023	3.1
2024	3.6
2025	4
2026	5.02

Will he meet his savings goal?

★ 3 Ed sets himself a savings goal: Each month he will deposit £1 more than the month previous. Ed deposits £1 on 1 January 2011. Using the interest rates in the table, find the amount Ed has in his account on 1 January 2020.

Dates	Annual effective rates of interest (%)
01 April 10–31 March 11	5
01 April 11–31 March 15	4.29
01 April 15–31 March 18	3
01 April 18–31 March 21	2.55

4 Peter has made an insurance claim. The insurance company will make 5 payments of £20 000 into his savings account every year for 5 years, always on 1 April. The first payment is received in 2025. If Peter's savings account has an annual, effective interest rate of 3.2%, what is the present value of the payments on 1 April 2025?

★ 5 Skye's granny opens a children's ISA when Skye is born. She pays in £1000 when she opens it, and pays in £400 every year on Skye's birthday until she is 18 years old. Her granny's final payment is on Skye's 18th birthday.

  a Assuming the ISA has an annual effective rate of interest of 1.5%, calculate:

   i the present value of the account when Skye is born

   ii the accumulated value on Skye's 18th birthday.

  b Arfaa's granny also opens an ISA with the same rate of interest when Arfaa is born. She also makes an initial deposit of £1000 and pays in money each year. But Arfaa's granny wishes to ensure that her yearly payments stay in line with inflation. When Arfaa turns one, her granny deposits £400, and then every year she increases the amount in line with inflation. Taking inflation to be 2.24% per year, calculate:

   i how much more money Arfaa will have on her 18th birthday compared with Skye.

When you have completed this exercise, check that you know how to print your answers in both regular view and formula view.

- I can use spreadsheets to carry out calculations in a way that another independent user can easily understand. ★ Exercise 13A Q3, 5
- I can identify when a spreadsheet is badly presented and suggest or make improvements to the presentation. ★ Exercise 13A Q7
- I can use spreadsheets to find the accumulated value of a series of payments ★ Exercise 13B Q3
- I can use spreadsheets to find the present value of a series of payments. ★ Exercise 13B Q5

14 Loan Schedules and Savings Goals

This chapter will show you how to:

- create a loan schedule
- use a spreadsheet to help you save towards a specific savings goal
- use and apply the Goal Seek function in the context of loan schedules and savings goals.

You should already know:

- how to implement recurrence relations in a spreadsheet
- how to use spreadsheets to find the accumulated value of a series of payments
- how to use spreadsheets to find the present value of a series of payments.

Loan schedules and savings goals

The amount of money that a person borrows is called the capital. Borrowers must pay back both the amount borrowed (the capital) and pay **interest** too. A **loan schedule** shows the payments to be made for a loan, breaking the payments down into **capital** and **interest**. Loan schedules help the customer and loan provider know exactly how much capital the customer still owes, and are used to calculate how much a customer should pay each month. A common way of paying back a loan is to make **level, monthly repayments**. This means that the customer pays the same amount of money each month, paying towards the capital and interest, over the course of the loan.

The **effective rate of interest** is the percentage rate of interest for a specific period of time. An effective rate of 4% per year means that for every £1 you borrow, you will pay 4 pence in interest after 1 year.

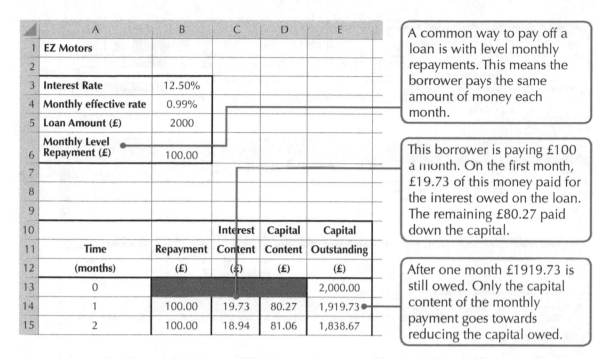

	A	B	C	D	E
1	EZ Motors				
2					
3	Interest Rate	12.50%			
4	Monthly effective rate	0.99%			
5	Loan Amount (£)	2000			
6	Monthly Level Repayment (£)	100.00			
7					
8					
9					
10			Interest	Capital	Capital
11	Time	Repayment	Content	Content	Outstanding
12	(months)	(£)	(£)	(£)	(£)
13	0				2,000.00
14	1	100.00	19.73	80.27	1,919.73
15	2	100.00	18.94	81.06	1,838.67

A common way to pay off a loan is with level monthly repayments. This means the borrower pays the same amount of money each month.

This borrower is paying £100 a month. On the first month, £19.73 of this money paid for the interest owed on the loan. The remaining £80.27 paid down the capital.

After one month £1919.73 is still owed. Only the capital content of the monthly payment goes towards reducing the capital owed.

Goal Seek

Goal Seek is helpful spreadsheet tool, it is found in the "Data" tab of Excel. Consider a simple example:

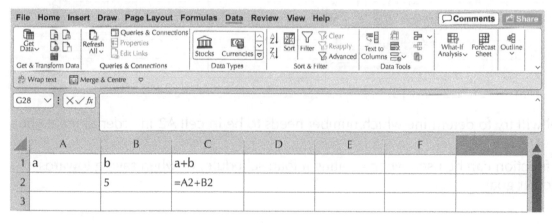

If you wanted to find out what number you should type into cell A2 to make cell C2 change to "42", you would use the following instructions:

1 Select cell C2.
2 Click "What-If Analysis".
3 Type "42" into the "To Value" field.
4 Click on the "By Changing Cell" field and select cell A2.
5 Click "OK".

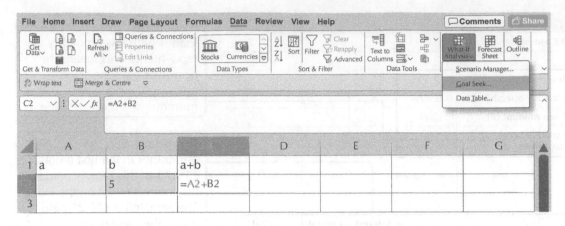

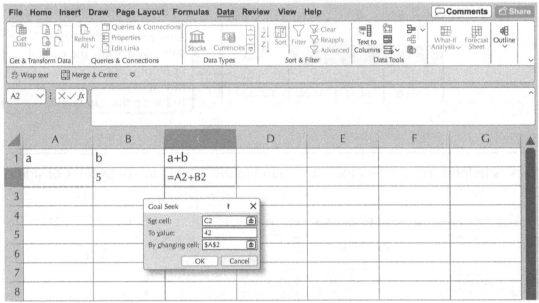

Excel will try to determine which number needs to be in cell A2 in order to meet your goal of "42".

This function can be used when creating a loan schedule, or when saving towards a particular goal.

Example 14.1

Declan borrows £7000 to pay for a car from EZ Motors. The annual effective interest rate is 7.1%. Declan will repay the loan in monthly, level repayments over 3 years. Calculate:

 a   The amount Declan will pay each month

 b   The value of Declan's final payment

 c   The total value of interest Declan will pay on his loan.

The spreadsheet is named "14.1".

First you must set up a loan schedule and work out the monthly effective rate. In the cell for your Monthly Level Repayment, you just enter a "dummy value". In this example, we will use £100.

> Using a dummy variable makes it easier to spot mistakes in your spreadsheet.

	A	B
1	EZ Motors	
2		
3	Interest Rate	7.10%
4	Monthly effective rate	0.57%
5	Loan Amount (£)	7000
6	Monthly Level Repayment (£)	100.00
7		

Next you must start creating the loan schedule.

	A	B	C	D	E
1	EZ Motors				
2					
3	Interest Rate	0.071			
4	Monthly effective rate	=(B3+1)^(1/12)-1			
5	Loan Amount (£)	7000			
6	Monthly Level Repayment (£)	100			
7					
8					
9					
10			Interest	Capital	Capital
11	Time	Repayment	Content	Content	Outstanding
12	(months)	(£)	(£)	(£)	(£)
13	0				=$B$5
14	1	=$B$6	=ROUND(E13*$B$4, 2)	=B14-C14	=E13-D14
15	2	=$B$6	=ROUND(E14*$B$4, 2)	=B15-C15	=E14-D15

The repayment column will reference the monthly level repayment. The interest content is found by multiplying the total outstanding on the loan by the monthly interest rate – this amount must be rounded. The capital content is the repayment amount with the interest content subtracted from it. Finally, the capital outstanding is the previous month's amount subtract the capital content from the current month.

As the loan is for 3 years, you then fill down the formulae until 36 months have passed.

(continued)

	Time	Repayment	Interest Content	Capital Content	Capital Outstanding
10					
11					
12	(months)	(£)	(£)	(£)	(£)
13	0				7,000.00
14	1	100.00	40.13	59.87	6,940.13
15	2	100.00	39.78	60.22	6,879.91
16	3	100.00	39.44	60.56	6,819.35
17	4	100.00	39.09	60.91	6,758.44
18	5	100.00	38.74	61.26	6,697.18
19	6	100.00	38.39	61.61	6,635.57
20	7	100.00	38.04	61.69	6,573.61
21	8	100.00	37.68	62.32	6,511.29
22	9	100.00	37.33	62.67	6,448.62
23	10	100.00	36.97	63.03	6,385.59
24	11	100.00	36.60	63.40	6,322.19
25	12	100.00	36.24	63.76	6,258.43
26	13	100.00	35.88	64.12	6,194.31
27	14	100.00	35.51	64.49	6,129.82
28	15	100.00	35.14	64.86	6,064.96
29	16	100.00	34.77	65.23	5,999.73
30	17	100.00	34.39	65.61	5,934.12
31	18	100.00	34.02	65.98	5,868.14
32	19	100.00	33.64	66.36	5,801.78
33	20	100.00	33.26	66.74	5,735.04
34	21	100.00	32.88	67.12	5,667.92
35	22	100.00	32.49	67.51	5,600.41
36	23	100.00	32.10	67.90	5,532.51
37	24	100.00	31.71	68.29	5,464.22
38	25	100.00	31.32	68.68	5,395.54
39	26	100.00	30.93	69.07	5,326.47
40	27	100.00	30.53	69.47	5,257.00
41	28	100.00	30.14	69.86	5,187.14
42	29	100.00	29.73	70.27	5,116.87
43	30	100.00	29.33	70.67	5,046.20
44	31	100.00	28.93	71.07	4,975.13
45	32	100.00	28.52	71.48	4,903.65
46	33	100.00	28.11	71.89	4,831.76
47	34	100.00	27.70	72.30	4,759.49
48	35	100.00	27.28	72.72	4,686.74
49	36	100.00	26.87	73.13	4,613.61

You can see that that the £100 dummy amount is much too small. If Declan only paid £100 a month, he would still owe £4613.61 at the end of his term.

You now use Goal Seek to find the correct value.

1 Select cell E49.

2 Click "What-If Analysis".

3 Type "0" into the "To Value" field.

4 Click on the "By Changing Cell" field and select cell B6.

5 Click "OK".

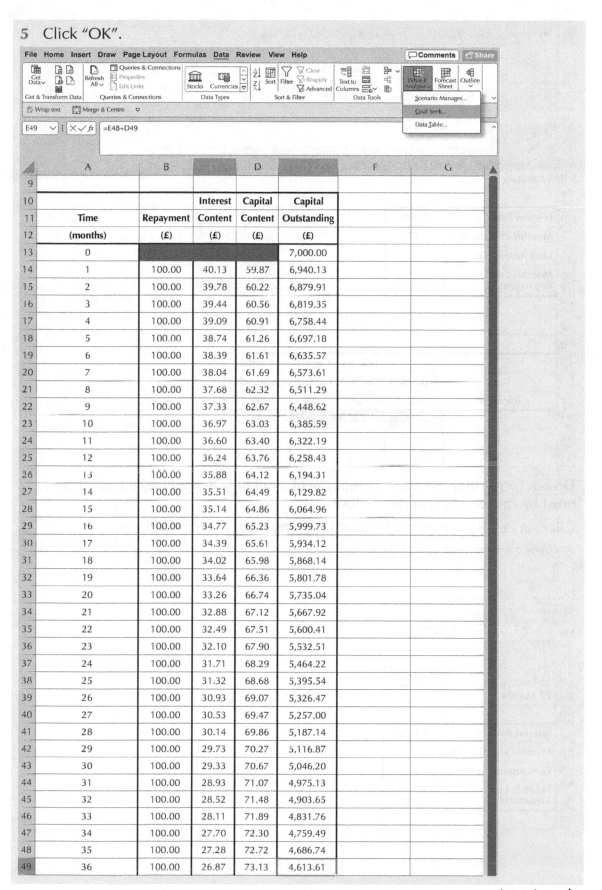

File | Home | Insert | Draw | Page Layout | Formulas | Data | Review | View | Help

Get & Transform Data — Queries & Connections — Data Types — Sort & Filter — Data Tools — What-If Analysis — Scenario Manager... — Goal Seek... — Data Table...

Wrap text   Merge & Centre

E49   =E48+D49

	A	B		D		F	G
	Time	Repayment	Interest Content	Capital Content	Capital Outstanding		
	(months)	(£)	(£)	(£)	(£)		
13	0				7,000.00		
14	1	100.00	40.13	59.87	6,940.13		
15	2	100.00	39.78	60.22	6,879.91		
16	3	100.00	39.44	60.56	6,819.35		
17	4	100.00	39.09	60.91	6,758.44		
18	5	100.00	38.74	61.26	6,697.18		
19	6	100.00	38.39	61.61	6,635.57		
20	7	100.00	38.04	61.69	6,573.61		
21	8	100.00	37.68	62.32	6,511.29		
22	9	100.00	37.33	62.67	6,448.62		
23	10	100.00	36.97	63.03	6,385.59		
24	11	100.00	36.60	63.40	6,322.19		
25	12	100.00	36.24	63.76	6,258.43		
26	13	100.00	35.88	64.12	6,194.31		
27	14	100.00	35.51	64.49	6,129.82		
28	15	100.00	35.14	64.86	6,064.96		
29	16	100.00	34.77	65.23	5,999.73		
30	17	100.00	34.39	65.61	5,934.12		
31	18	100.00	34.02	65.98	5,868.14		
32	19	100.00	33.64	66.36	5,801.78		
33	20	100.00	33.26	66.74	5,735.04		
34	21	100.00	32.88	67.12	5,667.92		
35	22	100.00	32.49	67.51	5,600.41		
36	23	100.00	32.10	67.90	5,532.51		
37	24	100.00	31.71	68.29	5,464.22		
38	25	100.00	31.32	68.68	5,395.54		
39	26	100.00	30.93	69.07	5,326.47		
40	27	100.00	30.53	69.47	5,257.00		
41	28	100.00	30.14	69.86	5,187.14		
42	29	100.00	29.73	70.27	5,116.87		
43	30	100.00	29.33	70.67	5,046.20		
44	31	100.00	28.93	71.07	4,975.13		
45	32	100.00	28.52	71.48	4,903.65		
46	33	100.00	28.11	71.89	4,831.76		
47	34	100.00	27.70	72.30	4,759.49		
48	35	100.00	27.28	72.72	4,686.74		
49	36	100.00	26.87	73.13	4,613.61		

(continued)

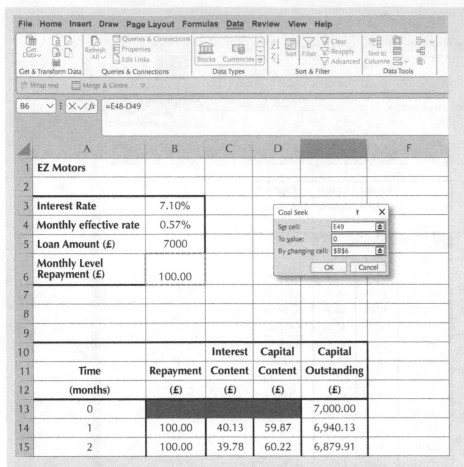

Do not forget that the monthly repayment amount is a monetary payment, and must be rounded to 2 decimal places.

Click on cell B6 and manually round the amount to 2 decimal places.

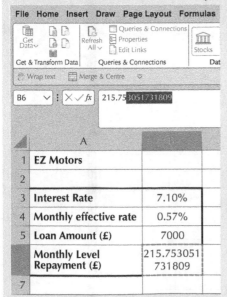

When you do this, you will find that the final amount of outstanding capital is no longer £0.

44	31	215.75	7.27	208.48	1,060.56
45	32	215.75	6.08	209.67	850.59
46	33	215.75	4.88	210.87	640.02
47	34	215.75	3.67	212.08	427.94
48	35	215.75	2.45	213.30	214.64
49	36	215.75	1.23	214.52	0.12
50					

12 pence is still outstanding after you round the level monthly payments.

You need to manually adjust the final payment. In this case, you will add 12 pence.

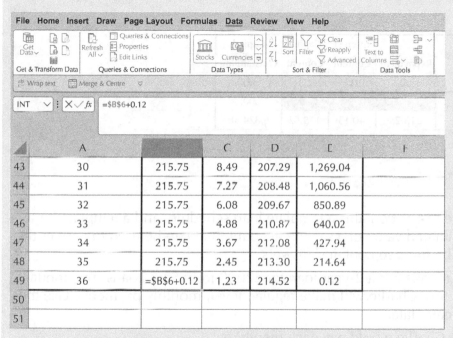

	A		C	D	E	F
43	30	215.75	8.49	207.29	1,269.04	
44	31	215.75	7.27	208.48	1,060.56	
45	32	215.75	6.08	209.67	850.89	
46	33	215.75	4.88	210.87	640.02	
47	34	215.75	3.67	212.08	427.94	
48	35	215.75	2.45	213.30	214.64	
49	36	=$B$6+0.12	1.23	214.52	0.12	
50						
51						

a Declan's monthly payments are £215.75.

b Declan's final monthly payment is £215.87.

c The total interest paid is found by summing the "Interest Content" column. The answer is £767.12.

(continued)

| E6 | ▾ | : | ✕ ✓ fx | =SUM(C13:C49) |

	A	B	C	D		F
1	EZ Motors					
2						
3	Interest Rate	7.10%				
4	Monthly effective rate	0.57%				
5	Loan Amount (£)	7000				
6	Monthly Level Repayment (£)	215.75		Total Interest	767.12	
7						
8						
9						
10			Interest	Capital	Capital	
11	Time	Repayment	Content	Content	Outstanding	
12	(months)	(£)	(£)	(£)	(£)	
13	0				7,000.00	
14	1	215.75	40.13	175.62	6,824.38	

Exercise 14A

⊕ The spreadsheet file "14A" contains a template loan schedule, and a template for a tracking the accumulated value of a savings account, that can be used to answer the below questions. The file also contains answers.

1 Charlie wishes to take out a 2-year loan for £2000 from Scot Loans. The annual effective rate is 5%. Charlie will make regular, level, monthly payments. Use the spreadsheet to calculate:

   a The amount Charlie would pay each month.

   b The amount Charlie would pay on his final month.

   c The total amount of interest Charlie would pay.

★ 2 Marley needs to save £5000 to pay for his wedding. He intends to deposit money at the start of every month, for 24 months. The effective rate of interest on his account is 3% for the first 12 months, 4.1% for the next 12 months. Marley wants to deposit the same amount of money each month. How much does he need to deposit each month to meet his savings goal?

3 Construct a loan schedule for a mortgage of £252 000 for 25 years, with an annual effective interest of 4%. The loan will be paid off in level, monthly payments. How much would the monthly payments be, and what would the final monthly payment be?

4 Sophie takes out a loan of £4000 to redecorate her house. The loan is to be paid back into level, monthly payments over 3 years. The effective rates of interest for the loan are detailed below:

Time	Effective rate
Year 1	3% per annum
Year 2	1.25% per quarter
Year 3	0.12% per month

  a Construct a loan schedule and find the value of each monthly repayment, and the final monthly repayment.

After making her payment on the 30th month, Sophie loses her job. She requests a 2-month payment holiday. Sophie will still need to pay off her loan within the 3 years.

  b Copy your spreadsheet and adjust it so that no payment is made on the 31st and 32nd month. Calculate the new value of Sophie's monthly payments for the remaining months.

> ⚠ Create a new **key variable** called "New Monthly Payment", and use Goal Seek again.

5 Mirren wishes to take out a loan of £3000 with an effective monthly interest rate of 0.6%. She will repay the loan in two years. She has three options for making repayments:

  • Make one payment at the end of 2 years.
  • Pay the interest each month, and then pay the capital at the end of 2 years.
  • Make regular, level monthly repayments.

For each option, calculate the total interest paid. Give one reason why Mirren may decide to choose the first option.

> ⚠ The first two options can be worked out on a calculator.

★ 6 Hannah is considering taking out a £4000 loan to pay for a car. The annual effective interest rate for the loan is 6.7%, and the loan would be paid back in level, monthly payments over 3 years.

  a Construct a loan schedule to show that the amount of the level monthly repayments would be £122.60. Find the amount of the final monthly payment.

Hannah reads the terms of the loan carefully and realises that the annual effective rate of interest will be raised to 9% per annum after 1 year. The monthly payments in the first year remain £122.60, but increase to a new level, monthly amount thereafter.

  b By adjusting your spreadsheet, calculate the amount of the level monthly repayments Hannah would pay after the first year.

★ 7 Oliwier has a savings account which pays 4.3% interest per year. He plans to deposit £200 into his account on 1 January 2022. Each month, on the first, he plans to make a deposit which is £2 greater than the month before. For example, he plans to deposit £202 on 1 February 2022.

    a Calculate what the total balance would be on 31 August 2022.

    b Calculate what the total amount of interest earnt would be on 31 December 2022.

    Oliwier wishes to save £4000 by the end of 2022.

    c Copy your spreadsheet and make suitable adjustments to find the initial payment Oliwier would have to make to reach this goal.

    Oliwier decides he wants to make an initial deposit of £200, and still reach his goal of £4000. To do this Oliwier will change how much he adds to each deposit each month. Currently, Oliwier's 'pay increment' is £2.

    d Copy your spreadsheet and make suitable adjustments to find the 'pay increment' Oliwier needs to meet this goal.

8 A high street bank offers a 25 year mortgage for £140 000. For the first two years the annual effective rate of interest rate is 2.52%, thereafter it becomes 3.99%. Calculate the amount of the level, monthly repayments for this mortgage.

9 Alice borrows £2887.34 from Cheshire Building Society. Alice will pay for the loan monthly, over the course of 100 months. The effective rate of interest is 0.5% per quarter. Each month's loan payment should be 1% greater than the previous month's payment. Calculate the total amount of interest Alice will pay for the loan.

10 A green energy company wishes to borrow £5 million to develop carbon capture technology. The money is borrowed with an effective rate of 8.55% per year. The loan is paid back in monthly instalments over 8 years. For the first 2 years, the company will only pay towards the interest content of the loan. For the next four years, the company will pay level, monthly amounts. For the final two years of the loan, the company will pay £100,000 per month.

    a Create a loan schedule for this loan. Include:

      i The level monthly payment amounts.

      ii The amount paid on the final month.

      iii The total interest paid.

    b Identify two risks to the lender in offering this loan.

- I can use create a loan schedule. ★ Exercise 14A Q5, 6
- I can use Goal Seek to save for a particular savings goal. ★ Exercise 14A Q2, 7
- I can use and apply the Goal Seek function. ★ Exercise 14A Q2, 6, 7

15 Salaries, Taxation and Financial Planning

This chapter will show you how to:

- understand taxation systems
- calculate tax on products purchased
- calculate salaries by calculating the component parts of the income and the deductions
- analyse and interpret the risks associated with financial planning.

You should already know:

- how to calculate the percentage of a quantity
- the difference between gross and net pay.

Understanding taxation systems

Governments collect taxes from their citizens to fund public expenditure. For example paying for hospital treatment, police, schools, public pensions and the armed forces. Taxation is taken in different ways: Income tax is applied to the money a person earns; Value Added Tax (VAT) is applied to certain purchases.

Value Added Tax (VAT)

VAT is paid on purchases in the UK. At the time of print, the VAT rates were as follows:

	% VAT	Purchases it applies to
Standard rate	20	Most goods and services
Reduced rate	5	Some goods and services, for example, home energy
Zero rate	0	Zero-rated goods, like children's clothing and most food

Example 15.1

Andreas spends £45.95 at the supermarket. £15.90 of his purchase was for items at the Zero rate of tax. The remaining items attracted the Standard rate of VAT.

What percentage of Andreas's supermarket spend was VAT?

Zero rate: He would have paid no VAT for the Zero rate items.

Standard Rate: £45.95 − £15.90 = £30.05. £30.05 of the total spend had a 20% VAT rate applied to it. You now reverse this percentage change to find the amount **before** VAT:

£30.05 ÷ 1.2 = £25.05.

This means the total VAT was £30.05 − £25.05 = £5.01.

The percentage of the supermarket spend that was VAT was $\frac{£5.01}{£45.95}$ = 10.9%.

Land and Building Transaction Tax (LBTT)

When buying property in Scotland, buyers must pay a land and buildings transaction tax (LBTT). The tax rate depends on the value of the property, whether the buyer is a first-time buyer, and whether the property is a second home for the buyer. The tax bands are set out in the table.

Band	Normal rate (%)	Second home rate (%)	First-time buyer rate (%)
Less than £145 000	0	4	0
£145 000–£175 000	2	6	0
£175 000–£250 000	2	6	2
£250 000–£325 000	5	9	5
£325 000–£750 000	10	14	10
Over £750 000	12	16	12

Example 15.2

Mark purchases a property for £275 000, paying LBTT at the normal rate. Calculate the total LBTT owed for this sale.

Band 1: Mark will pay no tax at all on the first £145 000 as the rate is 0%.

Band 2: Mark must pay 2% on the next £30 000 of the sale.
£175 000 − £145 000 = £30 000

2% × £30 000 = £600

Band 3: Mark must pay 2% on the next £75 000 of the sale.

2% × £175 000 = £1500
£250 000 − £175 000 = £75 000

Band 4: Mark must pay 5% on the final £25 000 of the sale.
£275 000 − £250 000 = £25 000

5% × £25 000 = £1250

The total tax paid is: £600 + £1500 + £1250 = £3350.

Exercise 15A

★ 1 Using the table given earlier for Land and Buildings Transactions Tax, calculate the LBTT for each of the following purchases:

  a Lewis is a first-time buyer. He purchases a home for £200 000.

  b Matthew is buying a second home for £350 000.

  c Lily is moving home. She has sold her property and has purchased a new home for £450 000.

  d Barbara is considering purchasing a property worth one million pounds. Calculate the LBTT for this purchase, assuming this would be using the second home rate.

  e A political party proposes changing LBTT so that any property sold over £750 000 would attract a flat rate tax of 15% of the sale price of the property. How much more tax would a buyer purchasing a £850 000 home pay if the policy was implemented? Assume the buyer would otherwise pay under the normal rate.

2 Louise goes clothes shopping and buys a mixture of adult and children's clothing. VAT for adult clothing is applied at the standard rate, and VAT for children's clothing is applied at the zero rate. Louise spent £145.95 in total. Of this, £39.50 was spent on children's clothes. Calculate the total amount of VAT Louise paid.

3 In 2012, the government proposed changing the VAT rate for certain hot foods, such as Cornish pasties, from the zero rate to the standard rate.

  If a pasty cost £2.20 before the proposed change, by how much would the price increase if the change was introduced?

★ 4 Dennis spends £85.60 at a shopping centre. £22.50 of his purchase was for items at the Zero rate of tax. The remaining items attracted the Standard rate of VAT. What percentage of Dennis's supermarket spend was VAT?

5 Kathleen's energy bill is £550 before VAT. Energy is taxed at the reduced rate. Next month her total energy bill will increase by 6%. Calculate her total bill for the following month.

Taxes on income

Income tax and **National Insurance** are forms of taxation. A percentage of a person's income is paid to the government. Many countries have a **progressive** tax rate, which means that the more a person earns, the higher their rate of tax. These different levels of tax are known as **tax bands**. In the UK the initial rate of income tax is 0%; this tax band is known as the **personal allowance**.

Below are the income tax bands for Scotland, for the financial year 2021–2022.

Band	Taxable income	Scottish tax rate (%)
Personal allowance	Up to £12 570	0
Starter rate	£12 570–£14 667	19
Basic rate	£14 667–£25 296	20
Intermediate rate	£25 296–£43 662	21
Higher rate	£43 662–£150 000	41
Top rate	Over £150 000	46

You begin paying National Insurance once you earn more than £184 a week (this is the amount for the 2021–22 tax year). The National Insurance rate you pay depends on how much you earn: 12% of your weekly earnings between £184 and £967; 2% of your weekly earnings above £967.

As well as taxes, most people pay pension contributions from their income. Pension contributions are made **before** paying income tax. National insurance is paid before pension contributions are made. Pensions exist to give you an income when you retire.

Example 15.3

Jemma is paid £9.18 an hour. She works 35 hours a week at this rate. Jemma is paid an additional 50% per hour when she works overtime.

In the 2021–22 financial year, Jemma worked 144 hours overtime. **Assume there are 52 weeks in a year.**

a Calculate Jemma's gross earnings for the year.

b Calculate Jemma's net earnings for the year, after paying income tax and National Insurance.

a The amount earnt at the basic rate will be £9.18 × 35 × 52 = £167 707.60.

The overtime pay is £9.18 × 1.5 × 144 = £1982.88.

Her total gross pay is £18 690.48.

b Her income tax is calculated by considering each band:

Jemma pays no tax on the first £12 570 of her pay.

Jemma must pay 19% tax on the next £2097 of her pay. This is 0.19 × £2097 = £398.43.

£14 667 − £12 570 = £2097.

Jemma's gross income is within the basic rate band. Her remaining pay of £4023.48 is taxed at 20%: 0.2 × £4023.48 = £804.70.

£18 690.48 − £14 667 = £4023.48.

Jemma's total income tax is (£0) + £398.43 + £804.70 = £1203.13.

(continued)

For National Insurance you first need to know how much Jemma earns a week:

£18 690.48 ÷ 52 = £359.43.

Again, consider each band:

Jemma pays no National Insurance of the first £184.

Jemma pays a 12% tax rate on the remaining £175.43: 0.12 × £175.43 = £21.0516.

£359.43 − £184 = £175.43.

Her annual National Insurance amount is 52 × £21.0516 = £1094.68.

Her total income tax and National Insurance amount is £2297.81.

Her net pay is £18 690.48 − £2297.81 = £16 392.67.

Example 15.4

Lynne starts a new job that pays £57 500 per year, plus a 6% bonus. Each month she contributes 3.5% of her monthly pay to her pension. Her employer also contributes to her pension, at 6.5% of her monthly pay.

Find the total amount of money added to her pension fund after Lynne has worked for 8 months.

Her total gross salary after the bonus is £57 500 × 1.06 = £60 950. This is £5079.17 per month.

Lynne's monthly pension contribution is £5079.17 × 0.035 = £177.77.

Her employer contributes £5079.17 × 0.065 = £330.15.

This means that the total monthly contribution is £177.77 + £330.15 = £507.92.

After 8 months this is £507.92 × 8 = £4063.36.

This sort of problem can also be solved using a spreadsheet. Open the spreadsheet file named "15.4" for the spreadsheet solution.

Exercise 15B

Refer to the 2021–22 income tax bands and National Insurance information to complete the following problems. For all calculations assume there are 52 weeks in a year.

1 Calculate the net pay of somebody earning £21 500 per year.

2 Elspeth says: "My current salary is £25 000. I want to turn down my 10% pay rise because if I accept it, I will move into a higher tax band and end up with less money per month." Explain why Elspeth is incorrect.

★ 3 Andy sells cars. Key features of his compensation include:

To use the current tax bands visit https://www.gov.uk/scottish-income-tax. For updated answers to this exercise, use the Income Tax Calculator spreadsheet.

- basic salary of £15 500
- a £50 bonus for every car sold.

Andy sells 123 cars this year. Calculate his net annual income.

4   Calculate the total **income tax** due for a salary of £55 000.

5   Danielle earns £2500 per month. In addition to this she earns 10% of her total sales for the month. If Danielle has sales of over £1000 in a month, she earns an additional payment of £150. Danielle contributes 5% of her monthly pay to her pension. Her employer will match her contributions.

In April, Danielle makes a total of £2850 in sales.

a   Calculate the amount of money that will be added to her pension for the month.

In May Danielle makes a total of £950 in sales.

b   Calculate the amount of money that will be added to her pension for the month.

6   Monty earns £98 000 per year. Monty says: "More than 50% of what I earn is taken in tax and National Insurance." Is this statement correct?

★ 7   Dylan has been offered two jobs, one in Edinburgh and one in London. Dylan currently lives in Edinburgh and has monthly expenses of £1300.

a   Dylan estimates that his monthly expenses will be 20% higher in London.

b   The Edinburgh job has an annual salary of £43 000.

c   The London job has an annual salary of £47 300.

d   If Dylan takes the London job, his income tax will be calculated using the English tax bands, given in the table.

Band	Taxable income	English tax rate (%)
Personal allowance	Up to £12 570	0
Basic rate	£12 570–£50 270	20
Higher rate	£50 270–£150 000	40
Additional rate	Over £150 000	45

For both jobs, calculate how much Dylan would have each month after considering National Insurance, income tax and monthly expenses. Comment on which job would leave Dylan with the most money left over each month.

8   For each of these salaries, calculate whether the earner would pay more income tax in England or in Scotland.

a   £12 500

b   £17 900

c   £24 750

d   £30 000

e   £27 393

⊕ 9   The spreadsheet file "15B" contains the August salaries for a sales team. As well as a flat rate bonus of 2% of their monthly salary, the employees receive a 8% bonus of their sales total for that month.

The employees contribute 3.5% of their total, gross, monthly pay to their pension. The company contributes 7% of the employees' total, gross, monthly pay to their pension as well.

Complete the spreadsheet and calculate the total amount the company contributes to employee pensions in August.

★ 10 Donald currently works 4 days per week part-time and earns a gross salary of £17 500 per year.

a Calculate how much Donald earns in a year in his current job after any tax and National Insurance contributions.

Donald pays £50 per day for childcare and £8 per day for travel.

Donald is entitled to 25 days annual leave when working full-time, or $\frac{4}{5}$ of these as a part time worker.

Donald does not need to pay for childcare and travel during annual leave.

Donald is paid in 12 equal monthly payments per year.

b Calculate how much Donald will earn per month, on average, after his costs are deducted.

c Donald is offered a full-time position, at 5 days a week, which pays a gross salary of £29 500. By how much money will Donald be better off per month if he takes this new job?

★ 11 Tamara works in sales. She earns a basic salary of £21 000 and earns a bonus of £2 for each sale she makes. On average she makes 17 sales per day.

Tamara works 4 days a week full time.

Tamara is entitled to 20 days annual leave.

Tamara does not earn any bonus when she is on annual leave.

a Calculate the total **income tax** Tamara pays in the 2021–22 financial year.

Tamara pays an average of £800 per month on her rent and bills.

Over the course of the year, Tamara pays £2212.32 in National Insurance contributions.

Tamara pays £15 a day to travel to work. She does not need to pay for travel during annual leave.

b Calculate how much Tamara earns per month, on average, after all her costs are deducted.

12 Below are historical tax bands from the year 2010–11:

Band	Taxable income	English tax rate (%)
Personal allowance	Up to £6475	0
Basic rate	£6475–£37 400	20
Higher rate	£37 400–£150 000	40
Additional rate	Over £150 000	45

For each of the following salaries, state in which year they would pay more income tax, 2010–11 or 2021–22.

a £8000

b £12 000

c £35 000

d £50 000

e £100 000

★ 13 Student loan repayments are only paid when somebody earns over a certain amount. This amount is called the threshold. Loan repayments are calculated at 9% of the amount earnt above this threshold.

The threshold in April 2021 was £2083 per month (before deductions).

Callum finishes university on 1 June 2021, with a total student loan of £20 460. He starts work at the start of August, earning £2917 per month, paid at the end of each month.

a Calculate the amount of income tax Callum would pay between August 2021 and April 2022.

Callum makes a single loan repayment to cover the period from 1 August to 31 March.

b Calculate the amount of this loan repayment.

The interest rate on the student loan is 0.01% per month.

c Calculate how much Callum owes on his student loan on 31 March 2022.

★ 14 Avery starts a job selling insurance. His company provides the following financial benefits **at the end of each month:**

> Salary: £1250 per month
> Commission: £12 for each insurance policy sold during the month
> Pension scheme: 3% of Avery's monthly earnings are paid into his pension fund.
> The company contributes a further 6% of the same monthly earnings.

Avery expects to sell 125 policies per month.

a Calculate how much income tax Avery expects to pay each year.

Avery's pension fund earns an annual effective rate of interest of 8%.

b Calculate the expected value of Avery's pension fund immediately after the third pension contribution is made.

> ⚠ Remember that Avery's pension contributions will be made before income tax is applied.

Analysing and interpreting the risks associated with financial planning

When planning for the future you need to be aware of what assumptions you are making. For example, the interest rate for a savings account or a loan may be subject to change. If the interest rate on your savings account drops, you may not save as much money as you had planned to.

When planning your finances you may need to make decisions about different financial products. For example, is it better to buy a car using your savings, or take out a loan? And what risks do you place yourself in either product?

Example 15.4

Charlie wishes to purchase a car that will cost £18000. He is considering two different options:

- Purchase the car outright using money from his savings account.
- Take out a bank loan for £18000 for a term of 36 months. Charlie would pay interest monthly and then pay off the capital at the end of the term of the loan. The effective rate of interest on the loan is 6.8% per annum, which is fixed for the length of the term.

Charlie currently has £18000 in a savings account that pays a variable rate of 6.4% per annum. This amount would remain in his savings account if Charlie took out the bank loan, and he would use his savings to make the final capital payment.

a   Charlie says: "If I take out the bank loan to fund the car purchase I'll be around £120 better off." Show by calculation that Charlie is correct.

b   What assumption has Charlie made?

Charlie decides to purchase the car through his savings.

c   Give two reasons Charlie may have decided to choose this option.

d   Give two risks of funding the car through his savings.

a   First you can calculate Charlie's bank balance if he left his savings in the account. £18000 × 1.064³ = £21681.90. In three years Charlie would accrue £3681.90 in interest.

Next you can consider the total cost of the loan. 6.8% per annum is:

$1.068^{\frac{1}{12}} - 1 = 0.55\%$, so each monthly payment would be

£18000 × 0.55% = £98.95.

The total interest paid would be 36 × £98.95 = £3562.20.

This means that if Charlie takes out the loan he will be able to make £3681.90 in interest through his savings, but will have to pay £3562.20 in total for the loan payments. The difference between these amounts is £119.70 so Charlie is correct.

b   Charlie has assumed that his savings account's interest rate will not change.

c   Charlie might believe the interest rate on his savings account will decrease, making the loan a better option. He may also wish to avoid having to pay money monthly.

d   Charlie is spending his entire savings to purchase the car. If he needs money in an emergency he will need to borrow money or be unable to pay. Another risk is the interest rate on his savings account might increase, which means he is losing out on more interest.

Exercise 15C

For this exercise refer to the spreadsheet file named "15C".

1 Seymour needs to save £650 in 2 years. His savings account has an effective rate of interest of 3.4%, which is fixed for 1 year. After a year, his interest rate will change by a maximum of 0.5 percentage points.

   Calculate the minimum amount Seymour needs to deposit to reach his goal.

★ 2 Audrey wishes to put money into a children's ISA for her daughter. She intends to deposit £10 at the beginning of each month, for every month until her daughter turns 16. Audrey makes the first deposit the month her daughter is born.

   The ISA pays an effective rate of 9.1% per year and Audrey assumes this will remain the same.

   a Complete the spreadsheet to calculate the total amount in the ISA when her daughter turns 16.

   When her daughter is born, Audrey receives a gift of £1500. Audrey decides to deposit the gift into the ISA at the beginning of the second month.

   b Copy your spreadsheet and adjust it to include the payment of the gift. How much more money will her daughter have when she turns 16 because of this deposit?

3 Bethany receives a gift of £4000 and is considering what to do with the money. She has a savings account that pays 1.25% per quarter.

   a Calculate how much interest Bethany would earn if she put her gift into her savings account for 20 months.

   Bethany has a personal loan. She has borrowed £10000 with an annual effective rate of 5.1%, to be paid back in level, monthly repayments over 30 months.

   b Complete the formulae in the loan schedule and calculate the level monthly repayment amount, and the final repayment amount.

   Bethany has just paid her 10th payment on the loan, and is considering using the £4000 gift to pay down the balance. The lender will recalculate a new level monthly amount for the remaining 20 payments.

   c Copy your spreadsheet and adjust it as required to calculate Bethany's new level monthly repayment.

   d How much money would Bethany save in interest payments by making this payment?

   e Should Bethany put her money into savings, or pay down her loan?

4 A farm shop wants to produce buffalo mozzarella. They intend to raise money by asking investors to invest £10000. Details about the scheme include:

   • Investors will receive £100 in gift vouchers, to spend at the farm shop, each month.

   • The investor will receive their £10000 back after 4 years.

- The investor's capital is at risk. If the venture fails, the farm shop will not be able to repay their investors.

  a What is the annual, effective rate of interest being paid to the investor?

  Sandra is deciding whether to invest in the farm shop or a fixed term bond. The fixed term bond pays an effective rate of interest of 0.5% per month.

  b What is the annual effective rate of interest of the fixed term bond?

  Sandra decides to invest in the farm shop.

  c Give two risks of investing in the farm shop.

★ 5 Mary decides to start saving regularly towards her retirement. She aims to retire from work on her 65th birthday.

Mary wants to estimate how much she will need to save by age 65 to cover her costs of living in retirement.

She expects these costs of living will be payable at the start of each month, from her 65th birthday, up to and including her 80th birthday. She estimates the costs will initially be £1300 at age 65 and will increase every month with inflation, at an effective rate of 2.25% per year.

Mary also expects that she will be able to earn an effective rate of interest of 4.5% per year on her savings during her retirement.

a Complete worksheet "5a" and fill in the relevant formulae in the to show that she must save approximately £200758.16 by her 65th birthday to cover her expected costs of living in retirement.

Mary has just celebrated her 20th birthday, and her monthly salary is £2600, which is constant and paid to her at the start of each month. She plans to make regular level contributions to her savings directly from her salary to meet her expected costs of living in retirement. She decides to make these contributions immediately when her salary is received, every month between now and age 65.

Mary expects to earn an effective rate of interest of 5% per year on her savings before retirement.

b Use worksheet "5b" to calculate what proportion of her salary she must save each month to meet her expected costs of living in retirement.

6 Jen is purchasing a car that costs £24730. She has two options available to her for financing the car:

Option 1 – A bank loan of £21000. The loan is for a term of 7 years. The annual effective rate of the loan is 2.9%. Jen would pay the loan in level, monthly payments. Jen would pay £3730 upfront for the car.

Option 2 – Finance through the car dealership. Key information about this finance:

- The length of the term is 48 months.
- The annual rate of interest is 5.9%.
- A £4000 deposit is needed for the car. Jen would have to pay £3000 towards this, but the car dealership would contribute £1000.
- Jen would make level, monthly payments throughout the term of the loan.

- At the end of the 48 months, Jen would pay a final payment of £10 000. This is referred to as a 'balloon payment'.

⬇ a Complete the worksheet "6a" to calculate the amount Jen would pay for her car if she took option 1.

⬇ b Complete the worksheet "6b" for the dealership loan. By how much cheaper is the bank loan than the dealership loan?

   c Give two reasons why Jen might decide to take the dealership loan.

- I can understand taxation systems.
- I can calculate tax on products purchased. ★ Exercise 15A Q1, 4
- I can calculate salaries by calculating the component parts of the income and the deductions. ★ Exercise 15B Q3, 7, 10, 11, 13, 14
- I can analyse and interpret the risks associated with financial planning. ★ Exercise 15C Q2, 5

---

16 PERT Charts and Gantt Charts

This chapter will show you how to:

- represent compound projects on Programme Evaluation and Review Technique (PERT) charts
- use systematic methods to find early and late times for activities, and identify critical activities and paths
- review and interpret PERT charts
- use Gantt charts to represent project activities
- compare Gantt and PERT charts.

You should already know:

- how to interpret a precedence table
- how to represent a precedence table in a diagram.

Activity networks

When managing a **project** some activities must **precede** others. For example, if a supermarket opens a new shop, they must first hire an architect to design the building before they start laying down foundations. Some activities in a project may take place at the same time. A supermarket can advertise vacancies and hire staff while builders are finishing construction of the shop.

Compound projects can be presented by activity networks, such as **Programme Evaluation and Review Technique** (PERT) charts. Activities can be either '**essential**' or '**critical**'. Essential activities must be completed to finish the project, but there may be flexibility in when they are completed. Critical activities are ones in which any delay in the activity would result in a delay in the project.

The "**critical path**" is the sequence of activities that allows the project to be completed in the shortest possible timescale.

Example 16.1

A supermarket owner wishes to complete a survey of some of his customers. Here is a precedence table showing a list of activities and the time taken to complete them:

Task	Detail	Preceding task	Time (hours)
A	Write survey	None	2
B	Recruit participants	None	1
C	Print surveys	Write survey	0.5
D	Carry out survey	Recruit participants, print surveys	2
E	Analyse results	Carry out survey	2.5
F	Produce a report	Analyse results	3
G	Send thank you letters to participants	Carry out survey	2.5

a Create an activity network for this project.

b State the critical path.

c What is the fastest possible time the project can be completed in?

d Which activities could be delayed without the project becoming delayed?

a The activity network could be laid out as follows:

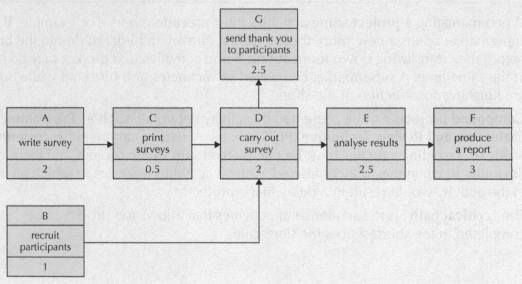

(continued)

b The critical path is ACDEF.

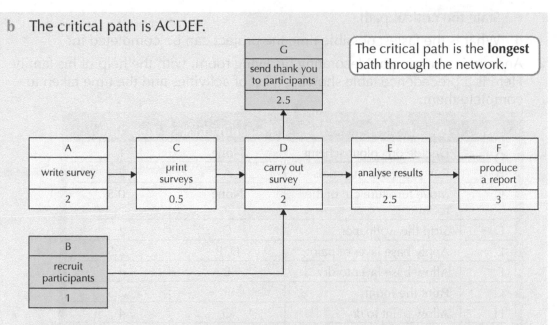

The critical path is the **longest** path through the network.

c The fastest time the project can be completed in is the sum of the times of the critical path: 2 + 0.5 + 2 + 2.5 + 3 = 10 hours.

d Activities B and G are essential activities, but not critical ones. B could be delayed by up to 1.5 hours without delaying the project, and G could be delayed by up to 3 hours.

The **fastest** time is found by following the **longest** path. This is because the project can only be finished as quickly as its longest path will allow.

The amount of time that a task can be delayed is called **'float'** or **'slack'** time.

Exercise 16A

1 Here is a precedence table for a project:

Task	Preceding task	Time (hours)
A	None	5
B	None	3
C	A, B	4
D	C	5
E	D	3

a Create an activity network for this project.

b Which activity is not critical?

   c State the critical path.

   d What is the fastest possible time the project can be completed in?

2 Andrew is planning on decorating his living room, with the help of his family. Here is a precedence table showing a list of activities and the time taken to complete them:

Task	Detail	Preceding task	Time (hours)
A	Decide on colour scheme	None	1
B	Buy paint and materials	A	2
C	Move furniture out of the room	None	0.5
D	Strip the wallpaper	C	2
E	Apply base layer of paint	D, B	2.5
F	Allow base layer to dry	E	4
G	Paint the room	F	2.5
H	Allow paint to dry	G	4
I	Tidy up	G	0.5
J	Put furniture back	I, H	0.5

   a Create an activity network for this job.

   b Which tasks are essential and which are critical?

   c State the critical path.

   d Andrew estimates that the job will take less than 17 hours. Is he correct?

★ 3 A company's managers are investigating how long it would take to launch a new product. They wish to launch before November 2024, which is in 18 months' time. The managers have created the following table of necessary tasks.

Task	Detail	Preceding task	Time (months)
A	Market research	None	3
B	Product design	None	8
C	Create prototype	B	3
D	Product testing	A, C	1
E	Analysis and pricing	D	1
F	Bug fixes	D	2
G	Staff training	E	3
H	Production	F, G	3
I	Marketing	E	2

   a Create an activity network and use it to calculate whether it would be possible to launch the product in time.

b A manager looks at the task table and makes the following corrections:

- Task I will now take 5 months
- Task B will now take 6 months.

Would this change your answer to part **a**?

★ 4 Here is an activity network:

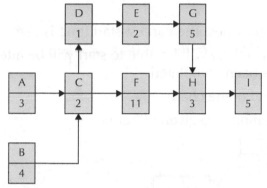

The duration of each activity is given in hours.

a Explain, using examples from this project, the difference between an activity that is essential for the project and an activity which is critical for the project.

b State the critical path for this project.

c What is the fastest time this project can be completed in.

d Explain, using an example from this project, what is meant by "slack time."

PERT Charts

Programme Evaluation and Review Technique (PERT) charts are a type of activity network. PERT charts are made up of **nodes** that contain the activity and its duration, as well as the earliest start time and the latest completion time for each activity.

Nodes contain each activity in the project. An activity will be something that needs to be completed in order for the project to be completed. For example, if your project was to launch a business, one activity may be to design your company's logo. Each **node** looks like this:

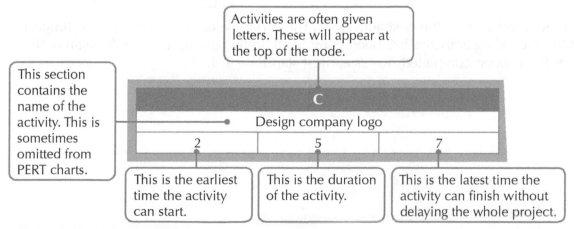

The advantage of using a PERT chart is so that it is easy to:

- identify the order of **precedence**
- identify the **critical path** and critical activities
- determine **slack time**.

The earliest start time for an activity can be found by completing a 'forward scan'. The process is:

1 Start with the activity with no preceding activities. The earliest start time is zero.

2 Move to the next activity. The earliest this activity will be able to start will be after the longest of the preceding activities has been completed.

3 Repeat this process until you reach the end of the chart.

Consider this activity network, where all durations are given in hours:

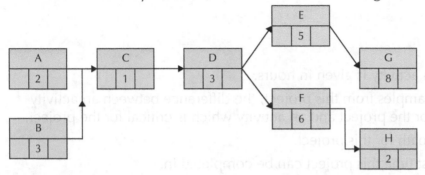

In a PERT chart, the number in the centre box is the duration. The left side box is the earliest start time. First, you put 0 as the earliest start time for the activities with no preceding activities, in this example activities A and B.

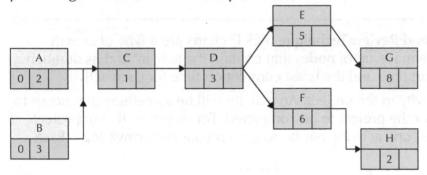

You now move on to the next activity, C. The earliest C can begin is after the **longest** of the preceding activities has been completed. In this example, C can't begin until after B has been completed, so C's earliest start time is 3.

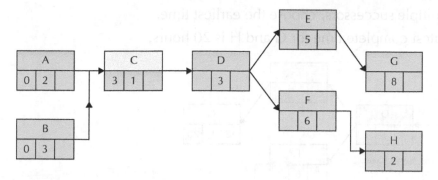

Next you move onto D. D can't begin until C is completed. The earliest C can be completed is after 3 + 1 = 4 hours. This is C's earliest start time plus C's duration.

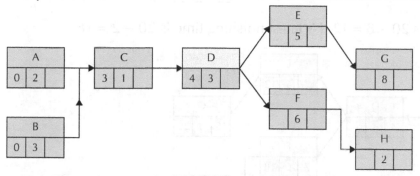

The earliest time D will be completed will be 4 + 3 = 7 hours. So the earliest start time is 7 hours for E and for F. Continuing the process you can complete the whole chart:

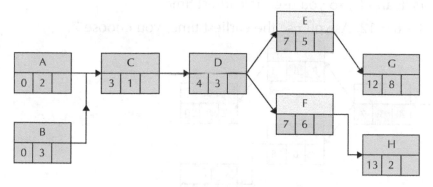

The empty boxes on the right of each node will be where you place the latest finishing times. To do this you need to first find the shortest duration the project can have. You can see the longest the path through the network is BCDEG, so 20 hours.

To complete a backwards scan:

1 Begin with the final activities in the network. Their latest completion time is the latest completion time of the whole project.

2 Subtract the duration from the latest completion time. Your answer is the latest completion time for the proceeding activity. Repeat this process until all activities have been calculated.

3 If an activity has multiple successors, choose the earliest time.

In your example, the latest complete time for G and H is 20 hours.

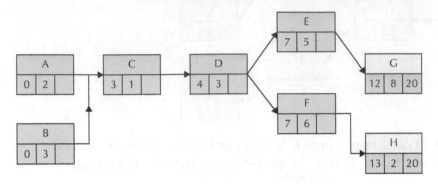

E's latest finishing time is $20 - 8 = 12$. F's latest finishing time is $20 - 2 = 18$.

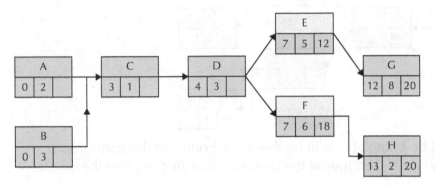

D has multiple successors (E and F), so you use the earliest time.

From E: $12 - 5 = 7$. F: $18 - 6 = 12$. As you use the earliest time, you choose 7.

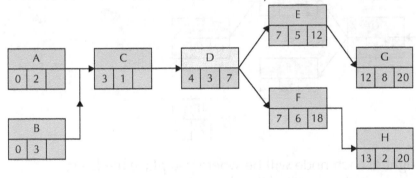

Continuing the process, you can complete the whole network:

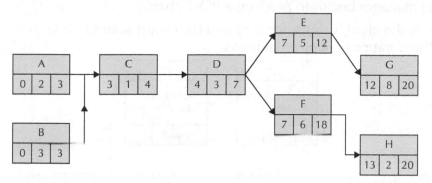

Notice that along the critical path BCDEG, the earliest start time plus the duration equals the latest completion time. This is not the case for activities that are only essential, in this example A, F and H.

The slack or float time can be found using the formula:

slack = latest finishing time − earliest start time − duration.

For example, F: $18 - 7 - 6 = 5$ hours. F has 5 hours of slack time, so could be delayed by up to 5 hours without delaying the project as a whole.

Exercise 16B

1 a Copy this activity network. Complete a forward and backward scan to fill in the missing values.

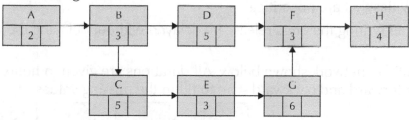

 b State the critical path.

 c Find the slack/float time for activity D.

2 A community centre wants to install an electric vehicle (EV) charging point at its premises. The project manager begins to produce a PERT chart.

a Copy and complete the chart, using a forward and backward scan to fill in the missing values. The durations are given in hours.

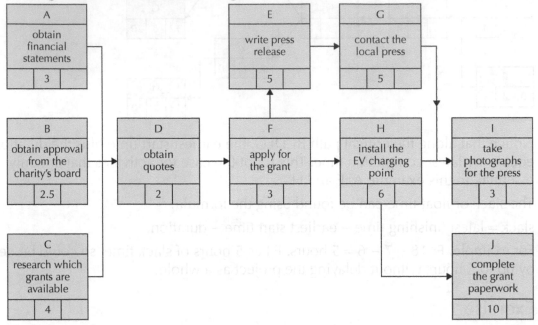

b State the critical path.

c List the activities that are not critical.

d How long can writing the press release be delayed without delaying the whole project?

3 a Copy the activity network shown below. All durations are given in hours. Complete a forward and backward scan to fill in the missing values.

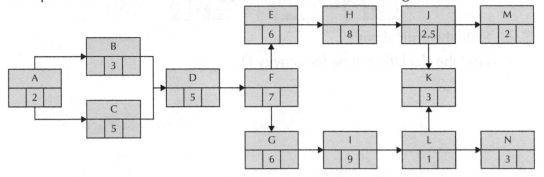

b State the critical path.

c If activity D was delayed by 4 hours, by how long would the whole project be delayed?

d If activity G was delayed by 2 hours, by how long would the whole project be delayed?

4   A group of game designers are taking part in a 'game jam', in which they must produce a computer game in a short period of time. Here is a precedence table showing the different tasks that must be completed.

Task	Detail	Preceding task	Duration (hours)
A	Write out ideas on paper	None	1
B	Code a simple version of the game	A	3
C	Produce music and sound effects	B	4
D	Produce artwork assets	B	5
E	Add levels to the game	B	7
F	Add music and artwork assets into the game	C D	1
G	Fix bugs	E F	4

a   Produce an activity network for this project.

b   Identify the critical path.

> If more than one critical paths exist, state both.

c   Complete a forward and backward scan to add the earliest start times and latest completion times to each activity node.

The team start work at 9:15 am. The game must be finished and submitted by 1:00 am. The team member working on music and sound effects is delayed by 1 hour.

d   Will the team be finished in time?

5   Here is a precedence table for a project.

Task	Preceding task	Time (months)
A	None	1
B	A	2
C	B	6
D	B	2
E	B	2.5
F	D, E	1
G	F	1
H	C, G, I	4
I	A	10
J	C, G, I	5

a   Produce an activity network for this project.

b   Identify the critical path.

c   Complete a forward and backward scan to add the earliest start times and latest completion times to each activity node.

★ 6　A political party is starting to campaign for an upcoming election in a council ward. Below is a list of activities that must be completed before voters receive their postal votes.

Task	Description	Preceding task	Time (days)
A	Write election leaflets	None	1
B	Contact volunteers	None	1
C	Have election leaflet printed	A	3
D	Deliver leaflet to target homes	B, C	5
E	Canvass target addresses	B, C	4
F	Phone voters	B	3
G	Promote hustings	None	1
H	Prepare candidate for hustings	G	2
I	Deliver second leaflet	D, E, F, H	5

a　Create an activity network for this project.

b　Write down the definition of 'critical path.' State the critical path for this project.

c　What is the latest possible time activity F can be completed to avoid delays to the whole project.

d　The party begins its campaigning on the 15 April. The voters receive their postal votes on the 31 April. Will the party complete its campaigning activities on time?

★ 7　A project manager produces a table for a project that is due to begin on Monday 3 January:

Task	Preceding task	Duration (working days)
A	None	3
B	A	7
C	None	9
D	B C E	3
E	None	1
F	D	2

a　Produce a PERT chart for the above project. Include the earliest start and latest finish times for each task.

b　On which date will the project be completed if there are no delays?

c　Calculate the slack time of activity C.

d　The project must be completed **before** Saturday 29 January. By how many days could activity C be delayed for and still reach this goal?

Gantt charts

Gantt charts were popularised by Henry Gantt in the 1910s. Gantt charts use bars to show the length of each activity and can also show the slack time. Activities are listed vertically, and their duration is shown horizontally.

To create a Gantt chart you need to know the duration, earliest start time, latest completion time and slack (or float) time. One way to do this is to first produce a PERT chart.

Activity								Day								
	1	2	3	4	5	6	7	8	9	10	11	12	13	14	15	16
A		■														
B			■	■	▨											
C			■	■	■											
D						■	■	■	■	■	■					
E												■				
F													■	■	■	■

A Gantt chart

The advantages of Gantt charts are:

- They are simple to understand.
- Each bar's length is relative to the length of time that activity takes, so it is easy to identify how long each activity takes.
- It is easy to see which activities can overlap.
- It is simple to find the critical path.
- The total time to complete the project is simple to calculate.

Example 16.2

Here is a Gantt chart for a project (note that float time is shown in a lighter colour):

Activity						Hours							
	1	2	3	4	5	6	7	8	9	10	11	12	13
A Strip wallpaper	■	■	■										
B Put down sheets	■	▨	▨										
C Apply first coat of paint				■	■								
D Let paint dry						■	■	■					
E Visit recycling centre				■	▨								
F Apply second coat of paint									■	■			
G Construct furniture				■	■	■	■	▨	▨				
H Move furniture into room											■	■	■

a State the critical path.

b How long has been planned to construct the furniture?

c State the float time for constructing the furniture.

d When is the earliest start time for visiting the recycling centre?

e When is the latest finishing time for visiting the recycling centre?

f What is the total duration of the project?

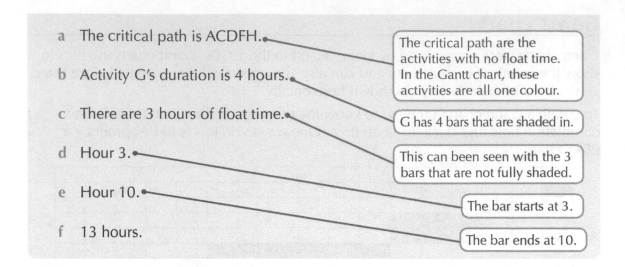

a   The critical path is ACDFH.

b   Activity G's duration is 4 hours.

c   There are 3 hours of float time.

d   Hour 3.

e   Hour 10.

f   13 hours.

The critical path are the activities with no float time. In the Gantt chart, these activities are all one colour.

G has 4 bars that are shaded in.

This can been seen with the 3 bars that are not fully shaded.

The bar starts at 3.

The bar ends at 10.

Example 16.3

Here is a precedence table for a project:

Task	Detail	Preceding task	Duration (months)
A	Market research	None	1
B	Develop adverts	A	3
C	Focus groups	B	2
D	Product development	None	6
E	Product testing	D	1
F	Project launch	C E	2

Produce a Gantt chart for this project, include float times within your chart.

In the exam, you may be given a grid in which to draw your Gantt chart. Alternatively, Gantt charts can be made in spreadsheet software. You will be advised whether to include float times.

(continued)

First sketch a PERT chart and find the earliest start and latest finishing times:

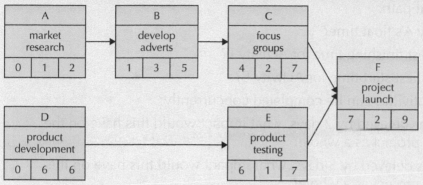

Next, work out the float times for each activity.

A = 1, B = 1, C = 1, D = 0, E = 0, F = 0.

Now you can create the Gantt chart. Each activity's bar will start from its earliest start time, and will be the length of its duration:

Activity	Month								
	1	2	3	4	5	6	7	8	9
A Market research									
B Develop adverts									
C Focus groups									
D Product development									
E Product testing									
F Project launch									

As you have been asked to include float times, you will add a lighter shade to fill in the float times to take each activity to its latest finishing time:

Activity	Month								
	1	2	3	4	5	6	7	8	9
A Market research									
B Develop adverts									
C Focus groups									
D Product development									
E Product testing									
F Project launch									

Exercise 16C

1   Here is a Gantt chart:

Activity	Day									
	1	2	3	4	5	6	7	8	9	10
A										
B										
C										
D										
E										
F										
G										

a   What is the shortest amount of time the project can be completed in?

b   State the critical path.

c   What is Activity A's float time?

d   What is the latest finishing time for Activity F?

e   What is the earliest start time for Activity B?

f   Which three Activities can be completed concurrently?

g   If Activity D was delayed by 2 days, what impact would this have on the duration of the project as a whole?

h   If Activity F was delayed by 3 days, what impact would this have on the duration of the project as a whole?

2   Below is a Gantt chart for a project being conducted by a market research team:

									Day					
Activity	1	2	3	4	5	6	7	8	9	10	11	12	13	14
A Write customer questionnaire														
B Collect data from customers														
C Input data into spreadsheet														
D Conduct focus groups														
E Questionnaire data analysis														
F Focus group data analysis														
G Write report														
H Share with client														

a   State the critical path.

b   What is the earliest time data can be inputted into the spreadsheet?

c   The project begins on Monday 4 April. On which date is it due to finish, assuming the project is worked on 7 days a week?

d   Activity F is delayed by 3 days. On which date would the project finish?

★ 3   A group of students are making a cartoon as part of an animation course. A PERT chart shows some activities for their project. The durations are measured in days.

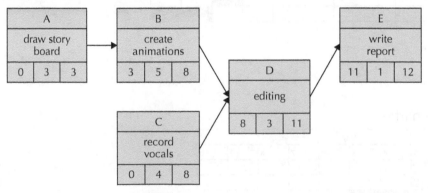

a   State the critical path.

b   Describe the three values in Activity D's node.

c   Produce a Gantt chart for the above network.

4 Below is a PERT chart for a project. All durations are given in hours.

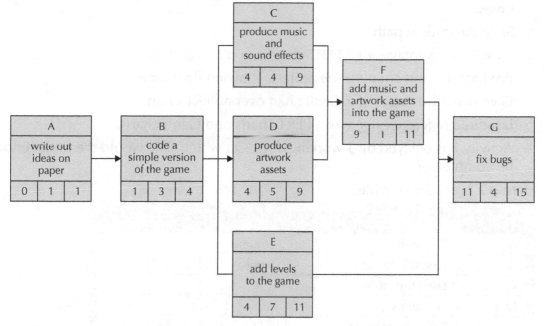

a Represent this project as a Gantt chart, showing float times.

b Which activities are essential, but not critical?

The project manager wishes to complete the project within 16 hours.

c Is this possible if there are no delays?

d If activity F was delayed by 1 hour, would this result in the project being delayed?

★ 5 This PERT chart shows part of a project for building a website. The durations given are in working days. The project begins on Monday 2 May.

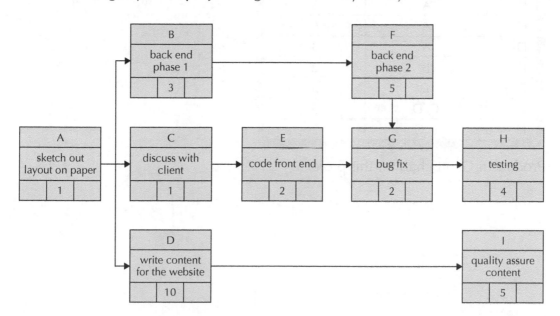

   a Copy and complete the PERT chart to show the earliest start and latest finishing times.

   b State the critical path.

   c Give two advantages of a PERT chart over a Gantt chart.

   d Produce a Gantt chart for the project, showing float times.

   e Give two advantages of a Gantt chart over a PERT chart.

   f List three differences between PERT charts and Gantt charts.

   g Activity F is delayed by 3 working days. On which date would the project now be completed?

6   Here is a precedence table:

Task	Detail	Preceding task	Time (hours)
A	Tidy hall	None	2
B	Unpack the van	None	1
C	Erect the tables	A	2
D	Set tables	B, C	2
E	Food preparation	D	4
F	Cook food	E	3
G	Put up decorations	D	5

Produce a Gantt chart for this project.

★ 7   A project as the following precedence table:

Task	Preceding task	Time (hours)
A	None	1
B	A	3
C	A	6
D	A	3
E	B	2
F	E	3
G	C, D	5
H	F, G	2

Produce a Gantt chart for this project.

8   Two flatmates are doing some housework. They have set out this precedence table:

Task	Description	Preceding task	Time (minutes)
A	Load the dishwasher	–	7
B	Run the dishwasher		80
C	Collect clothes for the washing machine	–	5
D	Load the washing machine		2
E	Run the washing machine		90
F	Empty the dishwasher		10
G	Empty the washing machine		3
H	Hang up clothes to dry		15

a   Copy and complete the precedence table, filling in the preceding task column.

b   Produce a PERT chart for the project, showing the earliest start times and latest finish times.

c   Produce a Gantt chart for the project, including float times.

d   The flatmates start the housework at 3pm. At what time can they expect to finish?

e   One of the flat mates is ill, so the other must complete all the activities themselves. Will this delay the finishing time?

• I can represent compound projects on Programme Evaluation and Review Technique (PERT) charts. ★ Exercise 16A Q3, 4

• I can use systematic methods to find early and late times for activities, and identify critical activities and paths. ★ Exercise 16B Q6, 7

• I can review and interpret PERT charts. ★ Exercise 16C Q3

• I can use Gantt charts to represent project activities. ★ Exercise 16C Q7

• I can compare Gantt and PERT charts. ★ Exercise 16C Q5

17 Risk and Expected Value

This chapter will show you how to:

- understand and interpret risk by calculating the expected value of costs and benefits of decisions
- apply expected value to real-life contexts
- identify the four main factors a project leader must consider for a project.

You should already know:

- how to solve simple probability problems
- how to determine the combinations of independent events.

Expected value

When considering a project, a **systems analyst** has four main factors to consider:

1 The reasons for initiating the project.

2 How feasible the project is.

3 Planning the project activities.

4 Controlling the project activities and the project team members.

Before starting a project, it is important to understand the **expected value** of the project.

Expected value = expected profits − expected costs

Expected profit is the probability of receiving a certain profit multiplied by that profit. The **expected cost** is the probability that a certain cost will be incurred multiplied by that cost.

Example 17.1

A company has been contracted to build a website for a client. If there is a delay in the website being launched, the company will be paid £9000 less. The website completion will be delayed in the event of staff illness. The company estimates that the probability of a delay is 5%.

A possible mitigation against this risk is that the company could hire extra staff for this project, at a cost of £2000.

a Calculate the expected value of costs that should be considered for the cost benefit analysis.

b Should the company hire extra staff for this project?

(continued)

There is a 5% chance of incurring a cost of £9000: $0.05 \times £9000 = £450$. As the cost of the mitigation is higher than the expected value of costs, the company should not hire the extra staff.

Example 17.2

Fiona buys a new phone for £850. Fiona is considering taking out mobile phone insurance. Her phone will need to be replaced in the event it is lost or broken by accidental damage. After conducting some research online, Fiona estimates that the probability she loses her phone in the next year is 2%, and the probability that she accidentally damages it is 6%.

Fiona has a choice of two insurance products:

Product 1 – Insurance that covers the loss of a phone. This costs £20 for the next year.

Product 2 – Insurance that covers both the loss of the phone and accidental damage resulting in the total loss of the phone. This costs £68 for the next year.

   a   Calculate the expected value of costs that should be considered for the cost benefit analysis.

   b   Calculate the expected value of costs if product 1 is purchased.

   c   Which product, if any, should Fiona purchase? Give a reason for your answer.

   d   Give two reasons why Fiona may decide to purchase product 2.

   a   First you should calculate the probability that phone is not lost or damaged.

        $(1 - 2\%) \times (1 - 6\%) = 92.12\%$

        So the probability that the phone is lost or damaged is $1 - 92.12\% = 7.88\%$.

> This is **not** the same as calculating $2\% + 6\% = 8\%$.

        The expected value of the cost would be $7.88\% \times £850 = £66.98$.

   b   Product 1 costs £20. When considering the expected value of costs for the phone, you only need to consider accidental damage, as loss would be covered by the insurance, $£850 \times 6\% = £51$. The total expected cost is $£20 + £51 = £71$.

   c   The expected value of costs for each insurance product is higher than the expected value of costs when not taking out insurance. So Fiona may decide not to purchase either product.

   d   Fiona might want the peace of mind of knowing that her phone is covered in the event of either type of loss. There might be other potential costs to losing her phone other than the cost of the phone itself; for example, if Fiona uses her phone for work, it would be a problem if she lost her phone and didn't have enough money to replace it.

Exercise 17A

1   A game at a school fête costs £2 to play. A spinner, as seen in the image, is spun. The spinner has equal sections made up of two yellow, two red, two blue and two green.

If the spinner lands on red, the player wins £6. If the spinner lands on any other section, the player loses.

Calculate the expected value of playing this game.

2   Paige is considering insuring her bike in the event it is stolen. Her bike cost £4000. After reviewing crime statistics for her area, she estimates that the probability that it is stolen in the next 5 years is 8%.

   a   Calculate the expected value of costs that should be considered for the cost–benefit analysis.

The insurance costs £300 for a 5-year term.

   b   Give one reason why Paige may decide to purchase the insurance.

3   Reece purchases a high-end gaming laptop for £4000. When purchasing the laptop, he is offered insurance that would replace the laptop in the event it stopped working within the first 5 years after purchase. Reece estimates the laptop has a 7% chance of breaking down within the first 5 years.

   a   Calculate the expected value of costs that should be considered for the cost benefit analysis.

The insurance costs £60 per year, for 5 years.

   b   Give one reason why Reece may decide to purchase the insurance.

   c   Give one reason why Reece may decide not to purchase the insurance.

Reece reads a review online that suggests that the laptop has around a 10% probability of breaking down within the first 5 years.

   d   Based on this information, should Reece purchase the insurance?

★ 4   Charlie is moving house and needs to rent a van. The cost of van rental is as follows:

   • £40 per day when booked in advance.
   • £55 per day if the van is needed longer than booked for.

Charlie definitely needs the van for two days. He estimates that there is a 30% probability that a delay will occur. In the event of a delay, he estimates that there would be 80% probability he would need the van one extra day, and a 20% probability that he would need the van two extra days.

Charlie has three options:

Option 1 – Book the van for 4 days.

Option 2 – Book the van for 3 days.

Option 3 – Book the van for 2 days.

a Calculate the expected cost of option 1.

b Calculate the probability that the van will need to be kept for one extra day.

c Calculate the probability that the van will need to be kept for two extra days.

d Calculate the expected cost of option 2.

e Calculate the expected cost of option 3.

f Based on the cost analysis you have completed, make a recommendation of which option Charlie should choose. Justify your answer.

> Remember that in the event of a delay of 2 days, the total additional cost will be £110.

★ 5 A group is organising its annual general meeting and needs to book a venue. A community centre hires its hall. The costs are as follows:

- If pre-booked, the cost is £50 for 3 hours, £60 for 4 hours or £65 for 5 hours.
- If not pre-booked, additional hours are charged at £20 per hour.

  The meeting is scheduled to take 2 hours. The group estimates that there is an 80% probability that the meeting will run late. If the meeting runs late, they estimate that there is a 50% probability it run an hour late, a 30% chance it runs 2 hours late and a 20% chance it runs 3 hours late. The group have three options:

  Option 1 – Book the room for 5 hours.

  Option 2 – Book the room for 4 hours.

  Option 3 – Book the room for 3 hours.

a Calculate the probability that the meeting will run on for 1 extra hour.

b Calculate the probability that the meeting will run on for 2 extra hours.

c Calculate the probability that the meeting will run on for 3 extra hours.

d Calculate the expected cost of option 2.

e Calculate the expected cost of option 3.

f Based on the cost analysis you have completed, make a recommendation of which option the group should choose. Justify your answer.

★ 6 A video games company intends to launch a new product before the Christmas shopping season. If the production of the game is delayed, the company expect to lose £300000 in sales. For the purposes of a cost-benefit analysis, it is assumed that there are two possible reasons the product may be delayed.

- A 0.25 probability of staff illness.
- A 0.05 probability of equipment failure.

a Calculate the expected value of costs that should be considered for the cost benefit analysis.

The company are considering two mitigations:

Measure A – Employ extra staff to replace anybody who falls unwell, at a cost of £10000.

Measure B – Inspect all the equipment, at a cost of £15000.

b Calculate the expected cost if Measure A is taken.

c Calculate the expected cost if Measure B is taken.

d Which, if any, mitigating measures should be taken?

7 A radio production company are recording a music documentary. If the production is delayed, the company will be charged an additional £5000. For the purposes of a cost-benefit analysis, it is assumed that there are two possible reasons for a delay:

- A 0.05 probability that equipment in the recording studio fails.
- A 0.01 probability of staff illness.

a Calculate the expected value of costs that should be considered for the cost benefit analysis.

The following control measures are considered:

Measure A – Book to place an additional studio on standby, at a cost of £120.

Measure B – Employ an extra member of staff to replace anybody who becomes unwell, at a cost of £40.

b Calculate the expected cost if Measure A is taken.

c Calculate the expected cost if Measure B is taken.

d Which, if any, mitigating measures should be taken?

8 A whisky producer is considering paying for an advertising campaign for their whisky. The advertising campaign would cost £400000. For the purposes of a cost-benefit analysis, the producer estimates there are three possible outcomes:

- A 0.6 probability the campaign is successful, and the producer makes an additional £900000 profit.
- A 0.2 probability that the campaign is a moderate success. The producer estimates that in this case they would make an additional £500000 profit.
- A 0.2 probability that the campaign is not successful, and there is no increase in profit.

The producer has the following options:

- Option 1 – Pay for the advertising campaign.
- Option 2 – Spend an additional £100000 on the campaign. In this event, the probability of the campaign being successful increases to 0.65, and the probability of it being a moderate success increases to 0.25.

- Option 3 – Decide to not go ahead with the advertising campaign.
a   Calculate the expected value of Option 1.
b   Calculate the expected value of Option 2.
c   Which option should be taken by the whisky producer? Justify your answer.

★ 9   A band wish to record an album. A local recording studio have the following costs:

- £350 per hour when pre-booked.
- If the session runs over the pre-booked time, there is a one-off fee of £50 and each additional hour costs £350.

   The band estimate the album should take 3 hours to record. For the purposes of a cost-benefit analysis the band assume there is only one reason a delay could occur, a 7% probability that there is a problem with the instruments.

   If a delay occurs, the band estimate there is a 80% probability an additional hour will be required, and a 20% probability that an additional two hours will be required.

   The band have three options:

   Option 1 – Book the studio for 5 hours.

   Option 2 – Book the studio for 4 hours.

   Option 3 – Book the studio for 3 hours.

a   Calculate the expected cost of option 1.
b   Calculate the probability that the band will need the studio for 4 hours.
c   Calculate the probability that the band will need the studio for 5 hours.
d   Calculate the expected cost of option 2.
e   Calculate the expected cost of option 3.

One control measure the band is considering is to pay to have their instruments inspected, at a cost of £50. They estimate that by doing this, the probability that a delay occurs is reduced to 1%.

f   Calculate the expected cost of option 2 if the control measure is taken.
g   Calculate the expected cost of option 3 if the control measure is taken.
h   What recommendation would you give to the band?

★ 10 An insurance company conducts a survey of its staff. A high proportion of staff complain that the software used to manage claims is too slow, leading to customer delays. Replacing the claims software would require a coordinated effort from the procurement team, the IT department, senior management and the claims teams. It is estimated that the expected cost of replacing the claims software would be approximately £250 000. The expected profit of replacing the software would is estimated as £450 000. A project leader is tasked with updating the claims software.

a   What are the four main factors the project leader should consider for this project?

b   Is this project feasible? Why or why not?

c   Give two reasons why the company may have decided to initiate this project.

d   Give one risk to pursuing this project.

11 A deck of cards is made up of 52 cards, 4 of which are Queens. At a charity fundraiser a game is played, the key details are as follows:

- The game costs £5 to play.
- The player draws a card from the shuffled deck. If the card is a Queen, the player wins £10, and may draw another card.
- If the player draws a second Queen, she wins £100, and may draw another card.
- If the player draws a third Queen, she wins £1000, and may draw another card.
- If the player draws a fourth Queen, she wins £10 000.
- The game ends when a player does not draw a Queen.

Calculate the expected value of playing this game.

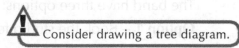 Consider drawing a tree diagram.

- I can understand and interpret risk by calculating the expected value of costs and benefits of decisions. ★ Exercise 17A Q4, 6, 9
- I can apply expected value to real-life contexts. ★ Exercise 17A Q4, 6, 9
- I can identify the four main factors a project leader must consider for a project. ★ Exercise 17A Q10

Appendix

This appendix contains:

- detailed examples of charts you can produce in RStudio
- common issues with R code or RStudio and how to fix them.

Charts in RStudio

All the R code that you need for the examination will be provided in the pre-release data booklet. This section will give you examples of more complicated diagrams, which may be of use to you when completing your project.

Each diagram has some example code that you type into RStudio and see the results.

Simple bar chart

Code: `barplot(table(X), main="Title", xlab="X Axis Label", ylab="Y Axis Label", las=2, names.arg=c("Bar 1", "Bar 2", "Bar 3"))`

Typing `las=2` in this function rotates the bar labels to a vertical position. Do not include this if you want the labels to be horizontal.

Example code:

```
chicken_weights <- chickwts
attach(chicken_weights)
head(chicken_weights)
table(feed)
barplot(table(feed), main="Barplot of Types of Chicken
Feed", xlab="Feed Type", ylab="Frequency", names.
arg=c("Casein", "Horsebean", "Linseed", "Meatmeal",
"Soybean", "Sunflower"), col=c("springgreen1", "thistle4",
"wheat3", "steelblue"))
```

Output:

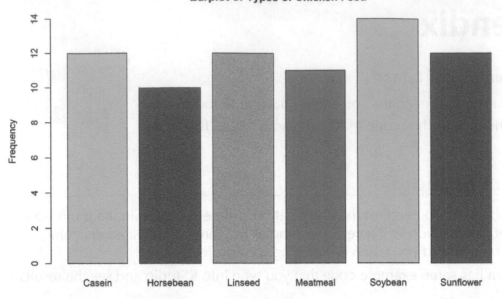

Barplot of Types of Chicken Feed

Stacked bar chart

Code: `barplot(table(X, Y), main="Title", xlab="X Axis Label", ylab= "Y Axis Label", las=2, names.arg=c("Bar 1", "Bar 2", "Bar 3"))`

`legend("LOCATION", legend=c("Label 1", "Label 2"), fill=c("gray1","gray50"))`

Replace LOCATION with any of: `top, bottom, left, right, topleft, topright, bottomleft, bottomright, center.`

Example code:

```
infertilitydata <- infert

attach(infertilitydata)

head(infertilitydata)

table(education, induced)

barplot(table(education, induced), main="Barplot of Number
of Induced Abortions and Education Level", xlab="Number of
Terminations", names.arg=c("None", "One", "Two or More"),
ylab="Frequency", col=c("lightskyblue3", "mediumblue",
"deeppink"))

##Add a legend

legend("topright", legend=c("0-5 years", "6-11 years", "12+
years"), fill=c("lightskyblue3", "mediumblue", "deeppink"))
```

Output:

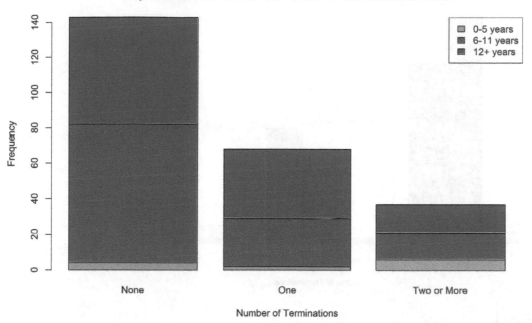

Side-by-side bar chart

Code: `barplot(table(X), main="Title", xlab="X Axis Label", ylab="Y Axis Label", las=2, names.arg=c("Bar 1", "Bar 2", "Bar 3"), beside=TRUE)`

`legend("LOCATION", legend=c("Label 1", "Label 2"), fill=c("gray1","gray50"))`

This code is identical to a stacked bar chart, with the addition of "besides=TRUE".

Example code:

```
infertilitydata <- infert
attach(infertilitydata)
head(infertilitydata)
table(education, induced)
barplot(table(education, induced), beside=TRUE,
main="Barplot of Number of Induced Abortions and Education
Level", xlab="Number of Terminations", names.arg=c("None",
"One", "Two or More"), ylab="Frequency",
col=c("lightskyblue3", "mediumblue", "deeppink"))
##Add a legend
legend("topright", legend=c("0-5 years", "6-11 years", "12+
years"), fill=c("lightskyblue3", "mediumblue", "deeppink"))
```

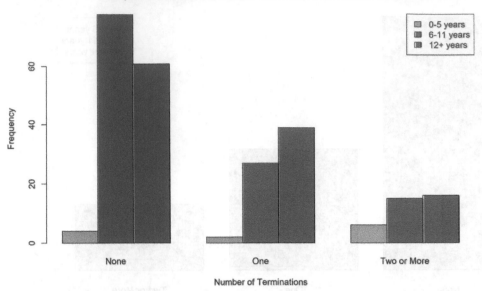

Pie chart

Code: `pie(table(X), main="title", col=c("gray1", "gray50", "gray90"), labels=c("Label 1", "Label 2", "Label 3"))`

You can add percentages to your pie chart using:

`percentages <- round(prop.table(table(X))*100, 0)`

`percentages_label <- paste(percentages, "%", sep="")`

`pie(table(X), main="title", col=c("gray1", "gray50", "gray90"), labels=percentagaes_label)`

Add a legend using:

`legend("LOCATION", legend=c("Label 1", "Label 2", "Label 3"), fill=c("gray1","gray50", "gray90"))`

Example code:

```
infertilitydata <- infert
attach(infertilitydata)
head(infertilitydata)
pie(table(education), main="Pie Chart Showing Education
Levels", col=c("cadetblue3","chocolate4","cyan4"),label
s=c("0-5 years", "6-11 years", "12+ years"))
### To add the % ages ###
percentages <- round(prop.table(table(education))*100, 0)
percentages_label <- paste(percentages, "%", sep="")
```

```
pie(table(education), main="Pie Chart Showing Education
Levels", col=rainbow(3),labels=percentages_label)
## Add a legend
legend("topleft", legend=c("0-5 years", "6-11 years", "12+
years"), fill=rainbow(3))
```

Output:

Pie Chart Showing Education Levels

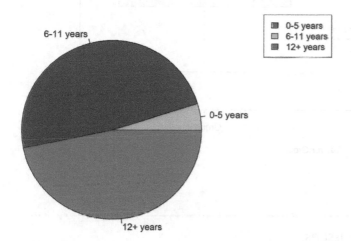

Pie Chart Showing Education Levels

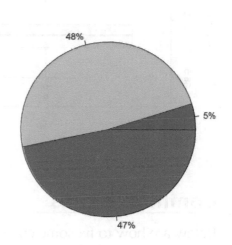

Multiple boxplots

Code: `boxplot(X, Y, main="title", xlab="X axis label", ylab="Y axis label", names=c("Name of first group", "Name of second group"), col=c("gray1", "gray50"))`

Example code:

```
beavers_data1 <- beaver1
attach(beavers_data1)
head(beavers_data1)
beavers_data2 <- beaver2
attach(beavers_data2)
head(beavers_data2)
boxplot(beavers_data1$temp, beavers_data2$temp,
main="Boxplot of Temperature in Beavers", xlab="Beaver
Group", names=c("Group 1", "Group 2"), ylab="Temperature
(Degrees Centigrade)", col=c("bisque", "burlywood2"))
```

Output:

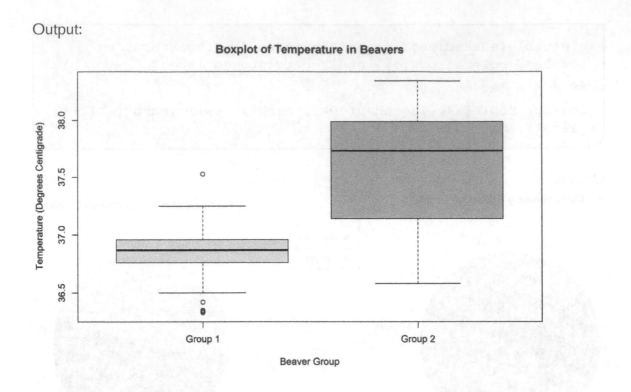

Boxplot of Temperature in Beavers

Common issues

Below are how to fix some common issues.

Importing data

If you are having issues importing data, follow these steps:

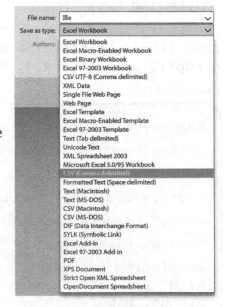

1 Check that your spreadsheet file is a CSV file. Click "Save as", and select the "CSV (Comma Delimited)" option. Do not select any other CSV option. A common mistake is to just name the file "file.CSV", but you must actually select the option in the drop down menu.

2 Press CTRL+SHIFT+H and make sure the working directory is set to the folder you saved the CSV file in.

3 Make sure the file name you type into RStudio matches the name of the file exactly.

RStudio's layout changes

If you accidentally close any of the windows within RStudio, you can quickly fix this by pressing CTRL+SHIFT+ALT+0.

If you change the size of the font in RStudio, this can be fixed by pressing CTRL+0.

Returning NA

When working with real data, it is common to see this error when finding summary statistics:

```
> mean(testdata)
[1] NA
```

This happens when a dataset has blank values, and the command can't calculate a numerical value.

To fix this add the following code: `na.rm=TRUE`.

```
> mean(testdata, na.rm = TRUE)
[1] 3.333333
```

Axis not fitting

When you produce a plot, sometimes the axes will not be long enough. For example:

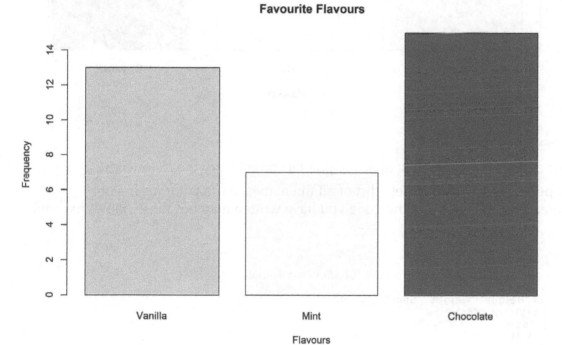

The bar for chocolate goes beyond the *y*-axis.

We can use the code **xlim and ylim** to set the upper and lower limits of the axes.

For example, **ylim=c(0,16)** will set a lower limit of 0 and an upper limit of 16.

```
ice cream <- c(1,1,1,1,1,1,1,1,1,2,2,2,2,2,2,2,3,3,3,3,3,3,3,3,3,3,3,3,3,3,1,1,1,1)
barplot(table(icecream),
        ylab="Frequency",
        main="Favourite Flavours",
        xlab="Flavours",
        names.arg = c("Vanilla", "Mint", "Chocolate"),
        col = c("wheat", "mintcream", "chocolate3"),
        ylim=c(0,16)
        )
```

Output:

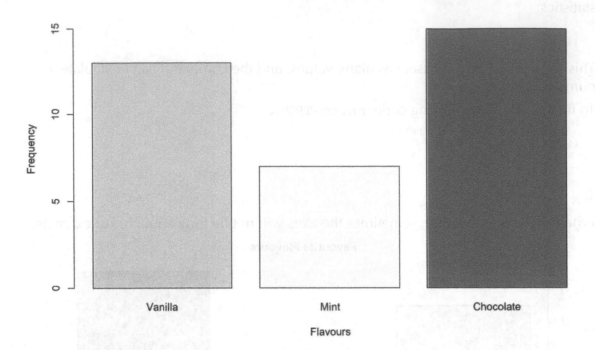

Other issues

If your code is not generating tables or graphs correctly, try the following:

1 Type names(mydata) to get a list of all the names of the columns in your spreadsheet. Check that the code you have written matches these names exactly.

For example:

```
> table(Gender)
Error in table(Gender) : object 'Gender' not found
> names(mydata)
[1] "Height" "Weight" "Shoe"  "gender"
> table(gender)
gender
    F M
31 6 6
```

Using names() can help you spot if you've included a capital letter when you shouldn't have, for example.

2   Open your spreadsheet, highlight the data you are using, and copy to a new spreadsheet. Import the new spreadsheet.

For example:

Copy old data:

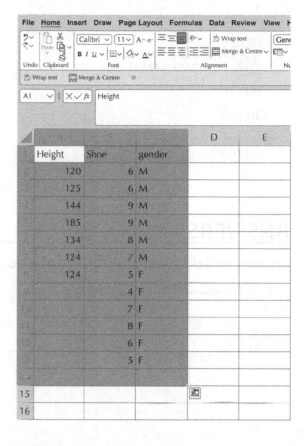

Paste into new spreadsheet:

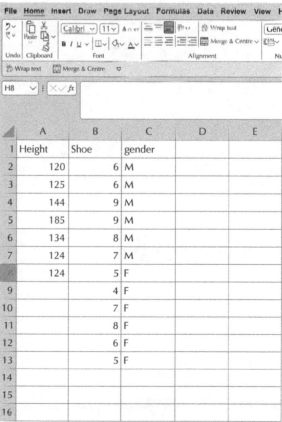

Import new data:

```
 8
 9  newdata <- read.csv("data2.csv")
10  attach(newdata)
11  table(gender)
12  |
```

Output:

```
> table(gender)
gender
F M
6 6
```

ANSWERS

Answers and downloadable files can be accessed at:

collins.co.uk/ScottishFreeResources/Maths